WRITING ACROSS THE MEDIA

Kristie Bunton
Thomas B. Connery
Stacey Frank Kanihan
Mark Neuzil
David Nimmer

University of St. Thomas

BEDFORD/ST. MARTIN'S
Boston/New York

For Bedford / St. Martin's
Developmental Editor: Simon Glick
Production Editor: Shuli Traub
Production Supervisor: Joe Ford
Marketing Manager: Charles Cavaliere
Art Director: Lucy Krikorian
Text and Cover Design: Paul Agresti
Copy Editor: Wendy Polhemus-Annibell
Composition: Pine Tree Composition, Inc.
Printing and Binding: Haddon Craftsmen, Inc.

President: Charles H. Christensen
Editorial Director: Joan E. Feinberg
Editor in Chief: Nancy Perry
Director of Editing, Design, and Production: Marcia Cohen
Managing Editor: Erica T. Appel

Library of Congress Catalog Card Number: 98-84336

Manufactured in the United States of America.

4   3   2   1
f   e   d   c   b

For information, write: Bedford/St. Martin's, 75 Arlington Street,
Boston, MA 02116   (617-426-7440)

ISBN: 0-312-15441-0

Acknowledgments

Acknowledgments and copyrights appear on pages 219–220, which constitute an
extension of the copyright page.

Preface

"Writing Across the Media" reflects the way we, the authors, teach media writing to mass communication students at the beginning of the 21st century. We emphasize storytelling, we integrate writing skills across media formats and we stress ethics. This book does, too.

The way we teach media writing, in turn, reflects changes both in mass media and in media writers' professional lives. We look at the World Wide Web, for instance, and we see that new media technologies offer ready opportunities for media writers who can adapt. We look at every mass medium from the Web to television to magazines, and we see that media writers' words have to work with visual or audio images in every instance. We look at the media writing that captures the attention and admiration of audiences and critics, and we see that the best media writers today relay complex information by telling articulate, engaging stories. We look at ethical lapses involving such national media as CNN and the Boston Globe, and such national advertisers as Calvin Klein and Holiday Inn, and we see that ethical responsibility is more important for media writers than ever.

This evidence persuaded us that today's students of media writing must learn basic storytelling skills that apply to multiple media formats and that are practiced with ethical sensitivity; we changed our media writing course accordingly. But when we looked for a textbook to use in our course, we could not find one that satisfied us. Most textbooks emphasized *print news* writing and tacked on chapters about *public relations* writing, *advertising copy* writing and *broadcast news* writing, but we wanted an integrated approach to media writing. Most textbooks emphasized reporting and information-gathering skills, not the storytelling skills we prize. Few textbooks considered ethical questions for media writers, and none addressed them with the detail we wanted. No textbook addressed the important marriage of visual or audio images with words that today's media writers must master.

Therefore, we conceived "Writing Across the Media," and we included all the features we sought. But we did not just write a textbook. We produced a complete teaching package, including a Video Guide and a World Wide Web site, that offers a flexible and interactive foundation for instruction.

The book can serve as the main text for introductory writing courses in various media. It can also be used as a supplementary text in introductory media writing classes, journalism writing classes or classes designed to introduce students in any major — English, communication, American studies — to media writing.

PRINCIPLES THAT GUIDE "WRITING ACROSS THE MEDIA"

This book's integrated approach to media writing is organized around three major principles:

Media Writers Tell Stories. The best media writers tell their audiences stories in a conversational style. Whether reporting on health care for a news magazine or describing a car for a television commercial, good media writers employ the language of literate conversation. They use crisp words, lively sentences and well-paced paragraphs to create word-pictures that show audiences their messages. These writers resist jargon, passive voice and stilted phrasing. Because we do, too, the tone we use in this book is at times less formal than that of a traditional text. We think of this book as a conversation about writing.

Media Writers Write for Multiple Formats. The best media writers can write in many ways. Media formats — broadcast commercials, radio newscasts, magazine articles or public relations news releases — are simply tools used by good media writers to tell stories. We instruct our students in basic writing skills that transfer from one format to another. This book stresses these skills, and it emphasizes the similarities and differences in applying these skills to different media formats.

Media Writers Engage in an Ethical Activity. The best media writers recognize and respect their power to bring audiences the world with every word they write. They wield that power fairly. Sensitive media writers exercise responsibility, select words carefully and resist clichés and stereotypes. We believe so strongly in media writing as an ethical activity that we devote the second chapter of our book to this topic, emphasizing ethics from the outset.

FEATURES OF "WRITING ACROSS THE MEDIA"

Focus on Writing. Because students will take other courses that focus on reporting, information gathering or research, we concentrate on writing at the introductory level. While we know that a piece of media writing is generally only as good as the information on which it is based, our focus in "Writing Across the Media" is on good, solid writing, especially the universal writing principles that are unfamiliar to so many of today's students. We do devote a chapter to strategies for information gathering, but we tie these strategies directly to the writing process.

Extensive Media Writing Examples. We believe *telling* students how to write is ineffective; they learn to write by writing, while we correct, coach and cheer. Beginning writers can, however, learn by imitating strong writing, so we have packed "Writing Across the Media" with colorful and current media writing examples. We selected these examples because they are well written. Additionally, they present a wide range of subjects, although we do not necessarily endorse all of their points of view. You may, in fact, disagree with some of these views, but we hope you appreciate the quality of the writing.

Two Integrated National Cases. To emphasize the idea that good media writers can write for multiple media formats once they have learned important basic skills, we include two integrated cases. These cases allow us to present a wealth of cross-media comparisons and written examples, in the form of news releases, advertisements and video scripts, that reinforce our integrated approach to media writing. An integrated approach makes sense because it emphasizes that the principles of good media writing apply to all media and it better prepares students for a world of mass media convergence.

The first case involves the announcement by Minneapolis-based Target Stores that it would not sell cigarettes in its 714 stores across the United States after Oct. 1, 1996. The second case involves a parental escort policy enacted on Oct. 4, 1996, by the largest enclosed shopping mall in the United States, Mall of America, in Bloomington, Minn. We picked these two cases because both topics — cigarettes and malls — involve consumer news that interests a broad segment of the population and resonates with students, thus leading to livelier class discussions about writing. By choosing two national cases based in our state, we were able to gather locally produced writing examples and compare them to nationally produced writing. We also included some examples of broadcast writing about the Mall of America case on the videotape accompanying our book; these video segments will enrich class discussion about examples presented in the text.

Practical Boxed Writing Tips. In each chapter of "Writing Across the Media," we present easy-to-find boxed writing tips that offer common-sense suggestions for beginning media writers. The tips cover "nuts and bolts" topics, such as using active voice and writing an inverted pyramid lead.

ORGANIZATION OF "WRITING ACROSS THE MEDIA"

The chapter organization in "Writing Across the Media" reflects the way we teach our media writing course. The first chapter gives an overview of the media writing process while Chapter 2 focuses on media writing as an ethical activity. Subsequent chapters offer a step-by-step guide to media writing, starting with basic skills and then moving on to more complex ones.

Chapter 1: Writing with Clarity and Coherence

This chapter explains the importance of good writing for all media writing formats and introduces the principles of good writing that will be used and illustrated throughout the book.

Chapter 2: Writing with Responsibility

Chapter 2 discusses media writing as an ethical activity. The discussion focuses on resolving conflicting values and loyalties, as well as the responsibility of telling the truth, avoiding harm and promoting social justice.

Chapter 3: Writing for Audiences

Although the principles of effective writing apply to all media formats, the specific medium and audience determine the tone, structure and content of the message. This chapter focuses on understanding various media and identifying audiences.

Chapter 4: Writing and Gathering Information

Chapter 4 presents the basics of information gathering, interviewing and observation, and considers the wealth of facts, opinions and documents available through the Internet, the World Wide Web and commercial database services.

Chapter 5: Writing the Opening

This chapter explains how to write both simple and sophisticated openings for articles, ads and broadcast segments. The chapter also explores the purpose of the opening paragraph and its many types.

Chapter 6: Writing Basic Stories

This chapter explains how to build from an opening to a short piece of copy by using the elements that make up the basic narrative form for media writing. It also shows how different media forms pose unique message-construction issues for the media writer.

Chapter 7: Writing with Visual and Audio Images

Chapter 7 explores the marriage of words and visual images, whether in broadcast scripts for video, advertising copy for photographs or copy blocks that accompany the color and graphics of the World Wide Web.

Chapter 8: Writing Complex Stories

The final chapter emphasizes that longer, more complex writing requires intelligent selection of material gathered from interviews, observation and other sources. More complex writing also depends on careful placement of that material, taking into consideration the ideas, themes and plots of the piece.

A NOTE ON ASSOCIATED PRESS STYLE

As a technical aside, we must point out that we have used Associated Press style throughout this book because beginning media writers need to master the standard stylistic rules used by working professionals. However, when the media writing examples we present in the text deviate from Associated Press rules, we retain the examples' original style.

ANCILLARIES

Two elements in addition to the textbook make "Writing Across the Media" a flexible package:

A Unique Video Guide. We offer instructors a unique videotape, along with an instructor's guide, to show students samples of colorful, energetic broadcast writing and to demonstrate that writing smoothly meshes words with pictures, natural sound and commentators. At the end of the video we also offer raw tape to let students put together their own stories, choosing sound bites and pictures to accompany their words.

The Video Guide is a dynamic tool for teaching media writing. First, its presentation of examples from the book moves discussion of writing from the page to the screen, where students can see a broadcast writer's words come to life. Examples from two media formats — broadcast news stories and a video news release — further reinforce the book's message about the importance of integrating media writing across formats. The inclusion of raw tape footage offers instructors the opportunity to integrate new writing components into their courses. This feature will be especially useful for faculty who have ignored writing for such formats as television news or broadcast commercials because they lacked access to high-quality raw footage.

An Interactive World Wide Web Site. The "Writing Across the Media" Web site is interactive and practical. It offers writing exercises, sample syllabi, teaching tips and current media writing examples. Placing these materials on the Web minimizes length and clutter in the book; it also allows us to update the book regularly. The Web site's interactivity will allow faculty to pick and choose from the pedagogical materials offered there, as well as to give us suggestions for additional materials.

ACKNOWLEDGMENTS

We know that nothing truly new can be said about being a good writer. Many people have spoken about this topic, and many people have exemplified it, including those who have taught us. Our goal with this book has been to put ideas about good writing into a different package, one that suits our needs and the needs of our students, and one that can be used by others who teach media writing.

We are passing on what we have learned from many, many sources and from our collective experience working in the media and teaching media writing. We owe much to those who were instrumental in developing the concept of a media writing course, particularly faculty at the universities of Oregon and Minnesota.

We are indebted to several people who have considerable reputations as teachers of writing. We have learned from their books and articles, but we have learned even more from their workshops, seminars and talks. Among these people are Roy Peter Clark and Don Fry of the Poynter Institute, Peter Jacobi of Indiana University, Paula LaRocque of the Dallas Morning News, the late Ed Orloff of the San Francisco Examiner and Don Ranly of the University of Missouri. Our files are packed with advice from them and many others, and their influence will be evident in this book.

In addition, we are indebted to many people who helped us with the details of producing this book. Among them are our own full-time and part-time faculty colleagues in the Department of Journalism and Mass Communication at the University of St. Thomas, where we were given generous personal support and helpful financial assistance. We thank Eric Larson, who provided computing support at crucial points. We also thank former students and professional colleagues who supplied some of our writing examples; they include Dixie Berg, Angela Keegan, Teresa McFarland, Joe McGrath, Quent Neufeld, Peter Noll, Terri Teuber, Amie Valentine and Peter Zapf. For their contributions to the videotape that accompanies this text, we thank our colleagues Karen Boros and Ron Riley, and we thank Ron Handberg, Tom Lindner and Jan McDaniel.

During our affiliation with Bedford/St. Martin's, we have been supported by several current or former staff members, including Laura Barthule, Simon Glick, Shuli Traub and Suzanne Phelps Weir. We thank them for their contributions. Finally, we thank the numerous faculty reviewers who helped us consider how to shape this project: David Boeyink, Indiana University; Anita Caldwell, Oklahoma State University; Jim Eiseman, Loyola University; Dana Eversole, Northeastern State University; Olan Farnall, Iowa State University; James Fields, University of Wisconsin; Jeffrey Finn, American University; Kathleen Hansen, University of Minnesota; Jack Lule, Lehigh University; Dianne Lynch, St. Michael's College; Jay Perkins, Louisiana State University; David Pritchard, University of Wisconsin–Milwaukee; Ford Risley, Penn State University; Jim St. Clair, Indiana University–Southeast; Edward J. Smith, Texas A&M University; Ken Waters, Pepperdine University; and Wendy S. Williams, American University.

Contents

About the Authors

The five authors of "Writing Across the Media" are faculty members in the Department of Journalism and Mass Communication at the University of St. Thomas in St. Paul, Minn., where each teaches, among other courses, Media Writing and Information Gathering. Their method of teaching that course was honored by the Teaching Standards Division of the Association for Education in Journalism and Mass Communication in 1997 and 1998 and was featured in the article, "Ability Grouping in Media Writing and the Gap in Mechanical Skills" by Kanihan, Bunton and Neuzil published in Journalism and Mass Communication Educator in 1998.

Kristie Bunton completed a Ph.D. in mass communication at Indiana University and master's and bachelor's degrees in journalism at the University of Missouri. She has worked as a newspaper reporter and in nonprofit public relations for public health and educational organizations. Her research, which examines the ethical performance of news media, has been published in the Journal of Mass Media Ethics. Bunton is a frequent commentator for Minnesota news media on issues of journalism ethics.

Thomas B. Connery has worked for various newspapers and the Associated Press and has written for magazines. He also has served as an adviser in nonprofit public relations. Connery is an authority on the history, nature and practice of literary journalism, and he edited "A Sourcebook of American Literary Journalism" (Greenwood, 1992). He received a Ph.D. in English from Brown University and an M.A. in journalism from Ohio State University.

Stacey Frank Kanihan holds a Ph.D. in communication from Stanford University, a master's in business administration from the University of South Florida and a bachelor's in English from Wellesley College. She teaches editing, public relations writing, mass communication research methods and advanced public relations. Her research focuses on political communication, and she has worked as a newspaper reporter, copy editor and corporate public relations writer.

Mark Neuzil is the co-author, with William Kovarik, of "Mass Media and Environmental Conflict: America's Green Crusades" (Sage, 1996). He has worked as a newspaper reporter and editor and wire-service editor, including stints for the Associated Press and the Star Tribune of Minneapolis-St. Paul. He is a member of the Society of Professional Journalists and the Society of Environmental Journalists. He received his Ph.D. in mass communication from the University of Minnesota.

David Nimmer is the author of two books of short stories and a 1993 winner of the Minnesota Book Award for personal narratives. He has worked as a reporter and managing editor at the Minneapolis Star, and as a reporter and associate news director at WCCO television, the CBS network affiliate station in Minneapolis-St. Paul. He holds a bachelor's from the University of Wisconsin. His teaching specialties are broadcast writing and reporting and newsroom management.

1
WRITING WITH CLARITY AND COHERENCE

*G*ood *writers are those who keep the language efficient. That is to say, keep it accurate, keep it clear.*
— Ezra Pound, "The ABC of Reading"[1]

As video, computer and telecommunication technology exploded through the 1980s, some observers predicted that printed or scripted prose used in both print and broadcast media would become mere window dressing for the visual, eventually becoming unnecessary or irrelevant. As we enter the twenty-first century, it hasn't happened.

In an age flush with visual images, Pound's notion of accurate, clear writing remains an integral component of our culture and society. Communication fuels our economic, political and social systems. Language — printed and spoken, simple or profound — shapes our national and personal identities, leads us to accept or reject values and beliefs, and allows us to define, construct and understand the world around us. Even with visual images, we need language to express, share and explain our joy and sadness, our love and hate, our ideas and emotions, our hopes and fears. Through language we give voice to our humanity, often by depicting our inhumanity.

The mass media are the major conveyors of our culture's messages, and their clear use of language remains necessary for effective communication, whether in a Nike ad or an organization's newsletter, in a news report or on a Web page. Media messages inform, persuade and entertain. They sell us products and ideas; tell us stories of heroes and villains; show us where to go, what to do and how to get there. Some messages are simple and direct: "Just Do It"; "Cloudy with a chance of showers today"; "The George Washington Bridge will be closed from midnight until 6 a.m."; "Police today charged a third suspect in the shooting and robbery at a Broadway jewelry store"; "Susan A. Jones, who began her High Tech Corp. career 20 years ago as a receptionist, has been named vice president of human resources." Such messages are efficient, giving us just the right amount of information.

Other media messages are more profound and far-reaching: a televised news report on the devastation caused by the Oklahoma bombing or a company newsletter story about the obstacles faced by a fellow worker struggling to raise a family and do her job well. In the following excerpt from a magazine article by Brit Robson, we get a firsthand account of his reaction to the premature birth of his son:

> Mingus emerged in a rush of purple, the red blood streaked like an oil slick over his blue body. His head seemed perilously soft, as pliable as a bean bag, and molded to an abnormally narrow size by the human canal he had just exited. He emitted a cry, tiny and fierce, that shook me like no other sound. The doctors had said Mingus would not be able to cry; indeed, the doctors had said he would not be able to breathe. His chance at life was estimated at one in twenty. But now, too suddenly for all concerned, I was a father. I was terrified that my son would die. I was more terrified that he would live.
>
> The facts speak for themselves. Mingus Paul Robson was born at 24 weeks gestation — four months premature. Due on June 10th, he was born February 17th. His weight was 1 pound, 9 ounces: the rough equivalent of six sticks of butter. At twelve and one quarter inches, he could fit on the cover of a record album. His eyes were fused together like a kitten's. And after that first spirited cry, he lapsed into a silent, painful, protracted struggle for his right to exist. It would be weeks before I heard his voice again.[2]

Similarly, CNN's Richard Blystone captures the poignancy and senselessness of a human tragedy in his description of the deliberate burning of oil fields in Kuwait. The piece begins with Blystone lighting a match in the dark, and then he says:

> Fire. The essence of violence. The war in Kuwait was over, but the violence was not. There was Saddam Hussein's parting curse.
>
> It seems as though if hell had a national park it would look like this. It doesn't say much for God's favorite creatures.
>
> One man decreed it, but many men methodically carried it out. None forbore what good could come of it, let alone what

harm: Stark, simple, monumental malice, with no possible benefit for anyone.³

Keep in mind that as Blystone speaks, video images of fire, smoke, people and animals regularly fill the screen, while his words interpret the images. Blystone goes on to tell us that "while the smoke sails off to paint a poison stripe across the globe, other men, the human powers, the government who mobilized hundreds of thousands to defeat Saddam Hussein, have mobilized a few dozen to defeat Saddam's curse ... and those who would pit their puny bodies and their crude machines against this rage of nature, have been waiting for equipment and supplies, fabricating what they can, working by firelight on lagoons to hold the millions of gallons of water they will need." One result of the fire, he says, is that "the soot insinuates itself into houses even with the windows sealed. Just keeping a little bit clean is an all-day, everyday job." Another result is that "brown birds are turning black, like the trees where they roost, like the air where they fly."⁴ Blystone's precise words and descriptive phrases give meaning to the video images, allowing viewers to grasp the significance of his report and to see and understand the consequences of the fire.

The preceding excerpts from Robson and Blystone, together with the many other examples appearing throughout "Writing Across the Media," demonstrate the central message of this book: that clear writing comes from careful writers, regardless of the media format. Although the purpose and audience of media writing may vary, in all media formats the need for clear, strong and effective writing remains constant. Consider yet another example, and as you read, think about where this writing might have appeared:

> Imagine if you could be a teenager in a generation other than your own. Would you wear bobby socks, circle skirts, varsity sweaters, bell-bottoms, tie-dye or grunge?
>
> Would you worship a crooner, a king or a rebel without a cause?
>
> Would you listen to swing, rock, disco or rap?
>
> And would your ticket say The Paramount, Woodstock or Lollapalooza?⁵

Is it from a newspaper feature on theme parties or dances, or an article on trendy fashions in a teen magazine? A radio commentary about how each generation of young people seeks its own identity? An item in a newsletter on cultural trends among young people? A column in Glamour, Vogue or another women's magazine? A news release announcing a special event at the mall? A story on teen styles over the ages in Time or Newsweek?

The copy could have been in any one of those sources, but it is from an advertisement for Seventeen magazine's 50th anniversary issue that appeared in Advertising Age. Regardless of the form or medium, then, all careful writers do the following:

- *Choose the right words and phrases*, including concrete nouns and vigorous verbs.
- *Make sentences flow*, by deleting needless words, using the active voice and varying sentence length to create rhythm and pacing.
- *Provide details*, including relevant information and focused observations.

Just as accomplished artisans know the names of the tools they use and how each tool works to help create or build an object, good writers know the tools of their craft — the parts of a sentence — and they understand how those tools are used to create clear, effective writing. They are wordsmiths who have mastered the vocabulary of their craft: noun, verb, subject, object, simile, metaphor, active and passive voice and so on. As you strive to do the same, keep in mind that memorizing definitions of these terms will not make you a good writer. But understanding how they fit in a sentence, what they do in a sentence and how they differ will allow you to distinguish between good sentences and bad sentences — and *that* will make you a better writer. Above all, you should aim to make your writing clear, because clarity means your message will get through to your readers.

ACHIEVING CLARITY

In "Pilgrim at Tinker Creek," writer Annie Dillard describes a scene from when she lived in the mountains in Virginia. As she walks along the edge of a creek scaring frogs, watching them jump from the bank into the water, she notices a small, green frog, half in and half out of the water. Here's how Dillard describes what happens next:

He didn't jump; I crept closer. At last I knelt on the island's winterkilled grass, lost, dumbstruck, staring at the frog in the creek just four feet away. He was a very small frog with wide, dull eyes. And just as I looked at him, he slowly crumpled and began to sag. The spirit vanished from his eyes as if snuffed. His skin emptied and drooped; his very skull seemed to collapse and settle like a kicked tent. He was shrinking before my eyes like a deflating football. I watched the taut, glistening skin on his shoulders ruck, and rumple, and fall. Soon, part of his skin, formless as a pricked balloon, lay in floating folds like bright scum on top of the water; it was a monstrous and terrifying thing. I gaped bewildered, appalled. An oval shadow hung in the water behind the drained frog; then the shadow glided away.

The frog skin bag started to sink.[6]

Dillard goes on to reveal that the shadow belongs to a giant water bug with strong, grasping forelegs, which "seizes a victim with these legs, hugs it tight, and paralyzes it with enzymes injected during a vicious bite."

Like Dillard, all writers face the challenge of description, including those who write for the mass media. Whether they write for newspapers or magazines, in newsletters or brochures, for radio or television or for the Internet or other multimedia sources, media writers must decide how to depict a scene or person or explain a topic, idea or product in a way that will be clear to their audience.

Although ambiguity can sometimes enrich writing, for most media writers achieving *clarity* is the primary objective. According to Eugene Roberts, the former editor at the Philadelphia Inquirer who shaped that paper into a "writer's newspaper," "There is no truer blueprint for successful writing than making your readers see. It is the essence of great writing."[7] Clarity allows us to *see* Dillard's frog-watching account. But how does she achieve it? She does not wave a magic wand or follow a prescribed formula for creating clear, vivid descriptions. Rather, she uses the same tools available to all writers, working with the most basic material: *words*. By choosing the right words, Dillard carefully crafts her picture.

Choosing the right words is an important first step toward achieving clarity. Clear writing also flows smoothly, includes appropriate details, and is aimed at a specific audience.

CHOOSING THE RIGHT WORDS

Words pinch, bite, cut and soothe; they shout, scream, boom and whisper. The careful writer cherishes both words and how they are used. Mark Twain said the difference between the right word and the almost-right word was the difference between lightning and a lightning bug. The careful writer thus strives for more lightning and fewer lightning bugs. As Eugene Roberts puts it, "The right words are to writing as the right tools are to carpentry."[8]

The right words are *concrete* and *specific*. They are strong nouns and vigorous verbs. Consider the number and type of vivid, active verbs Dillard uses to capture the frog-watching scene: *crumpled, sag, vanished, snuffed, emptied, drooped, collapse, settle, kicked, ruck, rumple, fall, glided, seizes, hugs, paralyzes.* Many of these verbs are short and sharp, some are one-syllable words and a few are uncommon (*ruck, rumple*). Although *crumpled, sag* and *drooped* are similar words, each verb has its own precise meaning, weight and purpose in Dillard's depiction. As Shakespeare has Hamlet say, "Suit the action to the word, the word to the action."

Of course, Dillard has the space to tell her tale and the time to shape it. Most media writers have less time and less space. An efficient one-sentence statement of Dillard's account might read this way: "I saw a giant water bug kill and eat a small, green frog" or "A giant water bug killed and ate a small, green frog." A stronger one-sentence description might say something like this: "I saw a giant water bug seize a small, green frog and suck out its life," or "A giant water bug seized a small, green frog and sucked out its life." Although *kill* is a strong verb, in this case it is less precise than *seize* and *suck*, which are concrete, active and colorful. So, even with less space, it is possible to achieve strong writing by using precise words.

Look at this description of Lanford Wilson's play "Burn This," from a magazine's theater listings:

> John Malkovich and Joan Allen co-star in a play by Lanford Wilson . . . about the romance between a married man and a dancer whose career up to now has consumed her, and whose present plight brings about major changes in her life.[9]

Nothing is "wrong" with this description. But it gives us an impression of the play as a somewhat familiar, even predictable drama, a

WRITING TIPS

USING THE TOOLS OF WRITING

Roy Peter Clark of the Poynter Institute for Media Studies is one of the most respected writing coaches in the country. At his workshops, Clark says it helps him to think of writing as similar to carpentry, complete with a plan and some 20 "writing tools" stored on a workbench. Here are several of Clark's tips for writing:

1. Begin sentences with subjects and verbs, letting subordinate elements branch off to the right. Even a very long sentence can be powerful when subject and verb make the sentence's meaning clear right from the start.
2. Place strong words at the beginning of sentences and paragraphs, and at the end. The period acts as a stop sign. Any word next to it gets noticed.
3. Avoid repeating a key word in a given sentence or paragraph, unless you intend a specific effect.
4. Prefer the simple over the technical: shorter words and paragraphs at the points of greatest complexity.
5. Slow the pace of information, for the sake of clarity. Short sentences make the reader move slowly. They give her time to think. They give him time to learn.
6. Reveal telling character traits. Don't say "enthusiastic" or "talkative," but create a scene in which the person reveals those characteristics to the reader.[1]

---

1. Roy Peter Clark, "If I Were a Carpenter: The Tools of the Writer," *Workbench* National Writers' Workshop (St. Petersburg, Fla.: Poynter Institute for Media Studies, 1994).

story we might regularly see on television. Now look at a different theater listing for the same play:

> Lanford Wilson's play can be considered an occasion for an astonishing performance by John Malkovich as a foulmouthed restaurant manager who woos and wins his dead brother's dancing partner.[10]

The concrete words and greater amount of detail in this second version make it stronger and livelier than the first version. We get the impression here that the play is something more than a routine drama, given the "astonishing performance" of one of the actors. Unlike the first version, this one uses vivid language: "married man" becomes "foul-mouthed restaurant manager," "dancer" becomes "dead brother's dancing partner," and the bland "romance between" them is now described in terms of "woos and wins," a punchy and alliterative alternative.

Much online writing also must pack vivid descriptions into small places. Microsoft's online "sidewalk" city guides, for instance, depend on brevity and punchiness to describe a wide range of activities and events in cities across the country. In a listing of children's activities at www.boston.sidewalk.com, the writers could have said, "Kids will see food dramatically and colorfully prepared right before their eyes." But in just two sentences, the online writers catch color and excitement through their choice of words:

> They don't need to be budding gourmets. They just need
> to like the flash of knives going chop-chop-chop and food
> being tossed and seared.[11]

Similarly, in describing Seattle's best holiday light displays, the writers at www.seattle.sidewalk.com describe Candy Cane Lane as a place where "giant animated reindeer and gingerbread men leap over the road, while a fly-fisherman casts his lure at passing vehicles" and "a giant 15-foot red teddy bear rests on his haunches, waiting for folks to assemble for a photo."[12] In a small amount of space, the active verbs *flash, chop, tossed, seared, leap, casts* and *rests* create concrete images in the reader's mind.

As the preceding examples suggest, effective writing often means choosing words that are plain and simple as well as exact. Look, for example, at the James Hardie ad shown in Figure 1-1, page 10. The ad promotes James Hardie Siding Products as having a long history of withstanding nature's elements with the headline "Frustrating Mother Nature since 1903." Below the headline is a photo of an attractive, well-maintained home that stands in a clearing. Dark, threatening clouds sweep over the mountains and woods behind the house. Here are the first three paragraphs of the ad's copy:

> We've seen torrential rains. We've seen howling winds.
> We've seen Mother Nature pitch some real fits.

Yet for nearly a hundred years, James Hardie building products have weathered them all. And you'll find that same stubborn reliability in our Hardiplank® siding.

Because Hardiplank siding won't rot, warp, buckle or swell. No matter what the climate.[13]

While there is nothing fancy or elaborate about this description, it is vivid and concrete, and it nicely plays off the frustrating Mother Nature image in the ad's headline. The opening paragraph uses three sentences with a basic noun-verb-subject construction that quickly lets the reader know what has been involved in that struggle with nature. The repetition of "we've seen" personalizes the product and makes dealing with the elements a concrete experience, something real rather than abstract. The next two paragraphs expand on that fight with nature. First, the copy claims that the James Hardie product has successfully resisted nature's onslaught, and then it lists the awful things that the James Hardie siding resists. The verbs are crisp and active: *pitch, weathered, find, rot, warp, buckle, swell.* The nouns are concrete, made more so by a limited use of modifiers: *torrential rains, howling winds.* The ad says much more than "When it rains . . ." or "When the wind blows . . ." Through careful word choice, the ad creates a picture of a product that withstands harsh weather and the damage it brings.

Using the right words does not require lots of space, nor does it necessarily require more time. Journalism is full of deadline writing that sparkles with well-chosen words and phrases. Compare, for instance, the following two leads (or openings) to a story about a volcano that erupted in Colombia in 1985:

BOGOTA, Colombia — A volcano that had been rumbling to life for months erupted, melting its snowcap and hurling down torrents of mud that buried four sleeping towns in an Andes mountain valley Thursday. Early estimates of the dead reached 20,000, and hundreds of bodies were being taken to a soccer stadium.

Blazing volcanic ash cascaded into the valleys Wednesday night. A few hours later the mud avalanche crashed through the towns, which had a combined population of 70,000.[14]

BOGOTA, Colombia — A volcano in central Colombia erupted last night, triggering floods in valleys below that buried large sections of at least two towns under tons of mud and rubble. Government and Red Cross officials said the death toll could reach 20,000.[15]

© 1997 James Hardie Building Products, Inc.

Figure 1-1

Both versions use words that create a picture in the reader's mind. In the first version, from the Associated Press, the volcano takes on the characteristics of a living, destructive being, "rumbling to life" and

"hurling down torrents of mud" at the towns. The second paragraph uses two contrasting verbs, *cascaded* and *crashed*, with dramatic effect. The second version, by Bradley Graham of The Washington Post,

creates a strong image by juxtaposing two specific subjects and verbs, *volcano . . . erupted* and *floods . . . buried*. But then Graham ends the sentence with the result of that action, vividly captured in straightforward yet alliterative prose: "two towns under tons of mud and rubble."

Graham started putting his article together on a plane to Bogota from the remote area where the volcano had hit, and he continued to work on it in a taxi from the airport to his office. When he got to the office at 4 p.m., he had three hours to write this story and a color story. Here's the beginning of the other story he filed on deadline. Notice how Graham creates suspense through foreshadowing with his ambiguous use of *it.*

> It came over them in the black of
> night, with the suddenness and force
> of a giant wave, swallowing everything
> in its wake.[16]

In the second paragraph, Graham tells readers that the "town drowned in a river of mud and stones," and residents spent "a night of horror in which family members were torn from each other as they struggled for air and secure ground." Graham talks with survivors "caked with mud, burned by the scalding temperature of the river of dirt and debris that overwhelmed their town, cut and battered from being dragged hundreds of yards across the valley floor."

The opening paragraph creates a sense of inevitable movement and destruction, emphasized by the image of a force "swallowing" whatever is in its way. Graham then builds on that image by carefully using strong, active verbs: *drowned, torn, struggled, caked, burned, cut, battered, dragged.*

Here is the lead to another breaking news story that paints a picture, this one by New York Times reporter Francis X. Clines:

> Three people were killed and dozens wounded today as an assailant threw grenades into a screaming crowd at a funeral and then fled across the graveyard from enraged mourners. Panic broke out and grieving families dived for cover by the mud of the open grave as four grenades exploded amid thousands of mourners gathered for the burials of three Irish Republican Army guerrillas.[17]

This is an example of a summary lead, one whose words create an image as well as convey facts (see Chapter 5 for more on leads). Clines says he deliberately used the word *graveyard* rather than *cemetery* because of the effect of "grenades in the graveyard" on sound and image.[18] We can see what Clines saw (he was there when it happened) through his choice of the specific verbs *killed, wounded, threw, fled* and *dived*, and his use of two well-placed adjectives, *screaming* and *enraged*.

Of course, words alone do not create clarity. In addition to choosing strong words, writers craft their depictions by carefully stringing their words together.

MAKING YOUR WRITING FLOW SMOOTHLY

A number of elements combine to make a piece of writing flow smoothly. As we have already seen, the most basic way to achieve flow and clarity is to write with verbs in the active voice, in sentences that have an actor or a subject doing something. The action must be expressed in the verb. For instance, consider again the sentence we gave earlier to sum up Dillard's frog-watching piece: "A giant water bug seized a small, green frog and sucked out its life." The bug is the actor, the seizing and sucking are the action and the frog is the object or the recipient of the action. That sentence is stronger, more dynamic than, say, "A small, green frog was seized and killed by a giant water bug," which is in the passive voice. It is passive because the subject is being acted upon; that is, the subject is the recipient rather than the doer of the action. Similarly, the ad for the Subaru Outback shown in Figure 1-2[19] would have been less effective had it been written in the passive voice. Instead of saying "First, it whipped the Blazer in a test of turning stability," and "Then it topped the Explorer in a test of braking," it could have passively declared that "First, the Blazer was whipped . . ." and "Then the Explorer was topped . . ."

The passive voice is usually more wordy than the active voice. Since another basic way to make your writing flow involves cutting out needless words, you should avoid passive constructions in most cases. Also avoid using too many adverbs and adjectives, which can slow down the pace and obscure the message, and omit irrelevant details, which clutter rather than clarify meaning.

Compare the Subaru ad copy on page 15, for instance, with this version:

> To begin with, the clumsy Blazer was whipped rather
> handily by the stylish and fine-handling Outback in a test of
> turning stability. In another test, the Explorer was topped

It outcorners. It outbrakes.
Heck, it even outsunroofs

Figure 1-2

by the dependable and reliable Outback in a test of braking. And finally, the Cherokee was beaten by the highly cost-efficient Outback through the Outback's superior fuel economy.

This example loses the punchy directness and power of the original. Replacing the original copy's focus on noun-verb action are vague,

imprecise modifiers (*stylish, dependable*), wordy, unnecessary phrases (*to begin with, in another test*), and redundancy (*highly cost-efficient*). The original moves much more naturally, pulling the reader easily along.

However, when modifiers are chosen with care, they can enrich description. For instance, in the Dillard passage quoted on page 5, "*slowly* crumpled" (rather than just "crumpled") emphasizes that the action does not occur in the blink of an eye. Later in that piece,

Dillard uses *bright* to describe how the scum looked. In both cases, the modifiers make the description more specific and move the sentences along smoothly. Similarly, when the Associated Press writer cited earlier writes "Blazing volcanic ash cascaded into the valleys," *blazing* adds significant detail, making the picture more vivid and the description more accurate.

Tightly constructed sentences in the active voice, with strong nouns and vigorous verbs, must naturally blend together. Not only must each sentence flow smoothly, but movement from one sentence to another should be smooth as well. Beginning writers, particularly media writers, often hear the admonition: Write short, with short words and short sentences. Because the beginning writer tends to write long sentences, cluttering them with too many words and ideas, "write short" is good advice. But the best advice is to write clearly. That often means using short words and short sentences, but not always.

The "write short" advice works well for short pieces, such as a few lines of ad copy or a few paragraphs in a publication, as well as in specific formats, such as a newsletter and parts of a newspaper. But writing composed of only short, simple sentences is monotonous and dull. Although such writing may sound less monotonous when it is spoken and voice inflections and pauses can adjust the pace, when longer writing repeats the same rhythm over and over, it soon stutters and sputters out.

Good writers thus change the pace of their sentences to keep the reader's attention. All of the writers we have looked at so far in this chapter vary sentence length for impact and emphasis as well as to engage their readers. In the following passage by John McPhee, for example, long sentences are used alongside one-word sentences. A one-word sentence is usually a sentence fragment and, technically, a grammatical error, but fragments can be used selectively by skilled writers to create emphasis and quicken the pace. McPhee worked on a farm as a way to get an insider's view of a New York City farmer's market for this article, which appeared in The New Yorker magazine. The paragraph comes just as he has finished placing boxes under a conveyer that fills the boxes with fresh-harvested onions:

> After thirty minutes of filling boxes, my arms feel as if they have gone eighteen innings each. I scarcely notice, though, under the dictates of the action, the complete concentration on the shifting of the crates, the hypnotic effect — veiling everything else in this black-surfaced hill-bordered surreally level world — of the cascade of golden onions. Onions. Onions. Multilayered, multilevelled, ovate, imbricated, white-fleshed, orange-scaled

onions. Native to Asia. Aromatic when bruised. When my turn is over and a break comes for me, I am so crazed with lust for these bulbous herbs — these enlarged, compressed buds — that I run to an unharvested row and pull from the earth a one-pound onion, rip off the membranous bulb coat, bare the flesh, and sink my teeth through leaf after leaf after savory mouth-needling sweet-sharp water-bearing leaf to the flowering stalk that is the center and the secret of the onion. Yash at the end of the day will give me three hundred pounds of onions to take home, and well past the fall they will stand in their sacks in a corner of the kitchen — the pluperfect preservers of sweet, fresh moisture — holding in winter the rains of summer.[20]

Obviously McPhee is having some fun writing sensually about onions. But notice how he uses sentence length to build intensity, with the extra-long sentence expressing the seemingly uncontrollable breathlessness of his desire. He also uses parallel constructions, such as in "*rip* off the membranous coat, *bare* the flesh, and *sink* my teeth. . . ."

Look again at the Dillard selection on page 5, noting how she varies sentence length for effect, to emphasize the action in what she sees and to focus the reader on each part of the scene as it unfolds. But very basically, Dillard follows the structure of the event — telling readers what she sees, as she sees it — and her account takes on the rhythm of her watching this act.

CNN's Richard Blystone is a master of language. He knows how to make words and visuals work together. Look again at his description of the burning Kuwaiti oil fields on pages 2–3. His reports are typically filled with striking word images that tell the viewers the significance and reality of what they are seeing in moving images. They also have almost perfect rhythm and pacing, as the words smoothly move viewers from idea to idea and image to image.

Here is another example from deadline news that shows how a writer can quickly establish a rhythm that matches the event covered. This is the first part of an article from the Middle East by Richard Ben Cramer, winner of a Pulitzer Prize for his reporting at the Philadelphia Inquirer. As you read the passage, try to identify the techniques and elements that allow it to flow smoothly:

NABATIYE, Lebanon — From the door of the shelter, the two men can see the town being blasted to bits, block by block. Houses, streets and shops disappear. When the smoke clears, there is only another square of garbage.

Shell after shell — perhaps 100 rounds — falls closer and closer, marching slowly up the hill toward the patio of the old stone villa under which the shelter was dug.

On the patio, near the shelter stairs, Sami is edgy. He takes a few steps, stops, looks toward the center of town, picks up his machine gun and puts it down again.

Sami is young, 21, he says, and he has been fighting for only three days. He is a Lebanese from the south, a lean, black-haired farm worker who took up arms only when the Israelis invaded Lebanon last week.

From the top step of the shelter, Shehabi, the other commando at the shelter door, tells Sami to get inside and settle down.

Shehabi is in his 30s. He carries a slight paunch as comfortably as he carries his rifle. He has been through many shellings in his six years with Al Fatah. "Take it easy," he tells Sami, "there's nothing you can do."

A far-off roar announces the approach of Israeli jets.

"Mirage," someone outside yells. "Everybody in. Down."

All scramble down the rough cement steps to the blackness below, in the heart of the hill.

The planes fly so high they are difficult to see. But the commandos insist that the air crews can see them on the ground if they move around in the open. So there are orders to stay underground, around the table littered with tea glasses and orange rinds, in the dank air that stinks of kerosene.

When the bombs hit, even though they are two-thirds of a mile away, the air in the shelter vibrates with a sound too low to hear. The glasses rattle. The talk stops.[21]

You probably noticed that the article moves with the description of the town being bombed, the bombing moving closer to the Lebanese commandos watching from their hiding place. But you also should have noticed the varied sentence length (in the last paragraph, especially), the strong verbs and specific nouns, the active voice and the parallel structure.

Cramer also uses with good effect the present tense and word repetition in this piece. For instance, instead of saying "Sami was edgy" or "He took few steps," he says "Sami *is* edgy" and "He *takes*," creating a sense of immediacy and moving the prose along. Instead of saying "Shells blast one block after another," Cramer uses repetition in "block by block" and "shell after shell." In addition, Cramer relies on alliteration, or the repetition of sound through similar sounding words, particularly words that begin with the same letter. He says, for example, "being blasted to bits, block by block." The fourth paragraph contains a series of *s* sounds: *shelter stairs, Sami, steps, stops, center.* Alliteration can help sear an image in the reader's mind, creating a melodic effect.

Cramer says that when he writes he is conscious of structure and rhythm. "I know how the sentence should sound to me," he says. "And events themselves have a rhythm. Moreover, stories have rhythm and if you change your rhythm badly or if you put in a jumble of different rhythms you lose something of the reader's interest and creative participation."[22]

Choosing the right words and making them flow are essential to the careful writer. But it is the job of the media writer to *get* the material. In all the examples we have looked at in this chapter, the writers had collected a wealth of details: facts and information, impressions and firsthand accounts.

USING APPROPRIATE DETAILS

Before media writers can actually begin writing, before they choose their words, fashion their phrases and string their sentences, they need to gather data, impressions and other appropriate details. They do this through observation and interviews, by scouring documents and marketing research (for advertising writers), by gathering background material and by gaining an understanding of the history and culture of their subject.

In a book about Newsweek magazine called "The World of Oz," Osborn Elliott tells a story about what it means to collect telling details:

> Frank Trippett, then a reporter at Newsweek's Atlanta bureau, recalled a classic phone conversation with an editor in New York.
>
> Trippett was in New Orleans in April 1962 when the editor telephoned. The Religion section, he said, was doing a story on the confrontation between Archbishop Joseph Rummel and an excommunicated parishioner named Mrs. Una Gailot over school desegregation. The confrontation had taken place on the archbishop's lawn and Emerson needed just two questions answered.
>
> "Number one," the editor said, "what does the archbishop's house look like? Is it wood, or stone, or brick? Is it Victorian with ivy on the walls?
>
> "What kind of day was it? Was it balmy and overcast, or hot and muggy?

"What does the archbishop look like? Is he old and be-spectacled or what? How did he walk when he came out of the house? Did he stride angrily? Or did he walk haltingly, leaning on a cane? How was he dressed?

"What is the walkway like? Is it concrete, brick or gravel? What do the grounds look like? Are there oak trees and rosebushes, magnolias and poppies? Were birds singing in the bushes?

"What was going on in the street outside the house? Was an angry crowd assembled? Or was there the normal business traffic, passing by oblivious to the drama inside?

"What were Mrs. Gailot and her friends wearing? Did they have on Sunday best or just casual clothes?

"What happened when the archbishop confronted Mrs. Gailot? Was he stern and silent? Or did he rebuke her? What was the exact language she used?

"Now," the editor said, "question number two. . . ."[23]

The New York editor urged Trippett to get the descriptive detail that would make the account lively, specific and colorful — that is, the type of information, based on observation, that shows readers what happened and allows them to experience the moment.

Good media writers have a keen eye for detail. They can pinpoint the look, smell, color and feel of what they experience or witness. In the passage by Richard Ben Cramer quoted earlier (pp. 17–18), for example, the writer tells us the Lebanese commandos have orders to stay underground and includes details that create a vivid image of the scene: ". . . there are orders to stay underground, around the table littered with tea glasses and orange rinds, in the dank air that stinks of kerosene."

Although media writers collect a wealth of information, they carefully select for inclusion in the piece of writing only those details that will add to the reader's understanding. Cramer tells us that Shehabi "carries a slight paunch as comfortably as he carries his rifle," a simple but revealing detail. It enables readers to see, but it does not bog down the narrative. Cramer could have described the walls, table, weapons and so forth. Instead, he sketches just enough of the picture so his readers can see, and smell, the commandos' dank, sparse hole in the ground.

All writers, not just Pulitzer Prize-winning ones like Cramer, can use vivid details to create a picture in the reader's mind. An agency in charge of public transportation in Minneapolis/St. Paul published in its employee newsletter a profile of a customer service representative. The writer's words quickly and efficiently allow readers to see the rep in action — all in a single sentence: "He carries his overstuffed day planner like a Bible and when he's on the street, bus drivers honk hello and customers wave from bus windows and from street corners."[24] We not only visualize the subject in action, but also see him as someone with a mission and who treats his job with religious fervor. Similarly, a news advisory from a university designed to call attention to the plight of the homeless might have simply declared that "During Homelessness Awareness Week students and faculty will sleep in makeshift quarters on the university quadrangle and hold prayer vigils." But the writer gives news and assignment editors a concrete image by writing instead that the students and faculty will be "sleeping in shanties made of cardboard and scrap lumber, and holding candlelight prayer vigils."[25] The words *cardboard, scrap lumber* and *candlelight* help to make the advisory more than a mere announcement of an event.

As good media writers go about selecting relevant and vivid details to use in their writing, they know that clichés should always be avoided. These hackneyed, overused expressions — such as "nipped in the bud," "black as coal," "warm as toast" and "as far as the eye can see" — are tired and dull rather than fresh and original. But what Donald Murray calls "clichés of vision"[26] also exist, and good writers avoid them, too. For instance, consider a few clichés of vision about cities and small towns: Cities are dangerous, noisy, filled with rude, uncaring people and unkempt, deranged men, whereas small-town America is safe, peaceful and filled with friendly folks, local characters and cracker-barrel philosophers. Of course, neither of these perspectives is necessarily true; rather, they are stereotypes of place, clichés of vision. Good writers see beyond and beneath such superficial views. (Chapter 2 examines the ethical implications of stereotypes in writing.)

However, creating fresh, original prose with vivid details involves much more than avoiding clichés, particularly in so-called "Gee whiz!" stories. For example, when a huge chunk of cliff broke free and roared down a hill and into a house in Fountain City, Wis., in 1995, regional reporters had a "Gee whiz!" story with striking visuals: the boulder in the center of the wrecked house, the path the boulder cut

WRITING TIPS
AVOIDING CLICHÉS

"Clear as a bell," "dead as a doornail," "backs to the wall," "smooth as silk," "the bottom line," "raining cats and dogs," "dress for success," "by hook or by crook," "light as a feather" — these are a few of the many familiar expressions people use in everyday conversation. Clichés are trite, overworked expressions that have lost their power and meaning over time, becoming vague and imprecise. Avoid using clichés in your writing. Save them — along with such contemporary expressions as "Been there, done that" and "Get a life!" — for family and friends.

Strive instead for originality in your writing. Challenge your imagination and try to come up with fresh words and phrases. As Lauren Kessler and Duncan McDonald advise: "So, as the sun sinks slowly in the western sky, be your own best friend and bid a fond farewell to the tried and true expressions that seem to creep into writing like a thief in the night, robbing it blind of its force."[1]

1. Lauren Kessler and Duncan McDonald, *When Words Collide*, 4th ed. (Belmont, Calif.: Wadsworth, 1996), 155.

through the trees, homeowners and residents unhurt but eager to talk about the event. But because stories like this one are relatively easy to write, they tend to be covered in a predictable way. The print reporter describes what happened, what damage was caused and how people reacted, then sprinkles the story with quotes from the homeowners and nearby residents. The lead may be clever or colorful. The television reporter does much the same thing, except more emphasis is put on pictures of the damage. Indeed, the images tell the story: The reporter stands in front of the wrecked house and boulder and looks amazed, says something about the unexpected "guest" dropping in for dinner and then fills the rest of the spot with comments from the homeowners and observers.

Sometimes a reporter in such a situation avoids the predictable kind of story. In the following example, TV reporter David Wildermuth incorporates words and pictures into an original account of the boulder incident in Wisconsin. As you read his story — preferably aloud, the way it was meant to be heard — notice the absence of clichéd expressions and clichés of vision and the preponderance of telling details:

(The piece opens with a video shot of a train whistle, and a train rolls across the screen.)

Live across from the tracks and you get to know the sound of an approaching freight train.

But at 440 North Shore Drive, the thunder normally knocks at the front door, not the back. 20 feet high, 30 feet wide, six feet thick. This was not one family's idea for an addition.

Lunchtime yesterday, the rock — the boulder — broke free from the cliffs above and . . . down it came.

(Comment from someone who saw boulder rolling.)

It's not like the boulder had a clear-cut path to the house. It cut its own, bulldozing trees *(camera quickly follows boulder's path, creating the sense of what it might have been like as the boulder tore through the trees).*

In some places the trench it left behind was five-feet deep. And it could have been worse. This part of the rock broke off. *(Reporter sits on rock.)*

(Comment from resident who remembers this happening 94 years ago, killing someone.)

True story. This time, though, no one was hurt, even though Maxine Anderson was inches away in her kitchen.

(Maxine Anderson talks.)

Sturdy as she looks, the house took such a jolt, the foundation's cracked, the doors are jammed and the locks won't lock.

Today, the valuables came out, the curious came in and the tourists drove by.

And the 80,000-pound center of attention? Far as we can tell, it isn't budging.[27]

Wildermuth's account gives viewers a considerable amount of information: about the sound of the boulder roaring down the hill, the location of the house, the size of the boulder, and the fact that the rock nearly hit someone. Note, too, how he tells the tale with concrete nouns, vigorous verbs and parallel constructions.

In advertising, collecting and using striking details are among the tasks involved in creating an effective ad. Occasionally, real people make it into an ad, and the ad copy reads much like a story in the news. For the most part, however, we tend to think of advertising copy writing as creative writing. Advertising writing does require creativity and imagination, as all writing does, but many print and broadcast ads for products, causes, activities and people incorporate purely functional writing, not unlike the most basic news story. An ad writer may rely on old formulas just as other media writers do because the formulas are necessary and they work; that is, they are functional. In addition, instead of being original and creative, instead of busting through clichés of vision, much advertising relies on common stereotypes.

The most effective ads are fresh and original, devoid of both tired formulas and stereotypes. But just as the creativity of journalists and public relations writers may be restricted by the need to use verifiable information, ad copywriters also must work within certain boundaries. They must structure an ad to suit the client, the client image, the product or brand image, the parameters of the marketing research (particularly the target audience) and the overall theme of the advertising campaign.

Thus, for example, the ad shown in Figure 1-3 was designed to suit the needs of Panasonic to sell its ShockWave portable audio products to a target market, identified as young people who think of themselves as rugged, take-charge individuals and who enjoy the challenge of testing their physical limits. The ad copy for the product must also reflect the campaign theme, "Take It to Extremes," which market researchers say will appeal to the target audience. The ad must fit the tone of the campaign as well as the tone of the publication in which it will appear, in this case Rolling Stone magazine. Notice that the ad copy runs under the headline "ShockWave, Rugged Audio Gear" and to the right of a dramatic photo. A rock climber hangs in the air at the top of a bent rock peak in the Black Hills, and in the middle of this image is "Take It to Extremes." Here is the ad's copy:

> The whole world is out there and you're gonna *leap, climb, tramp, tread, ride, plow, steam* right over it.
>
> Sounds rough? Not for the new *family* of Panasonic ShockWave portable audio gear. There's a ShockWave *CD player* with 10-second anti-skip memory and dual locking lid. A water-resistant, rubber-shelled ShockWave *personal stereo.* Plus an *AM/FM headset* with safety reflector and dual headbands for stability — all built for your rugged lifestyle.

Figure 1-3

Listen before you buy and decide for yourself. *You'll like what you hear*. After all, it is Panasonic sound.

ShockWave audio gear. Take it *higher, louder, faster*. Take it to extremes.[28]

The ad conveys information and attitude, details and style. As with all media writing, this ad is written for a specific audience.

WRITING FOR YOUR READERS

Media writers do not usually write for themselves or to satisfy their ego or muse. Almost always, media writers address a specific audience, and today that audience is less often a "mass" audience. Readers and viewers are more narrowly defined, perhaps by geography or interest, by income or gender. Blystone's report on the Kuwait oil fields (see pp. 2–3) is aimed at a broad national and international audience, but Cramer's story about the Lebanese commandos (pp. 17–18) is written for readers in the Philadelphia metropolitan area. Public relations writers may write for any number of audiences. For instance, sometimes a university relations office may have to create an article or brochure for several audiences — students, faculty, staff, neighbors, parents, alumni, the news media, legislators, donors. At other times, it may have to address only one of those groups, in which case the tone of an article written for faculty members would be different from that of an article geared to alumni.

The good media writer always keeps the audience in mind, understands what the audience needs and wants and tailors information and images to suit the specific audience. But for all audiences, the media writer strives for clarity because it is the most effective way to reach an audience. (We will talk more about writing for audiences in Chapter 3.)

THE NEXT STEP: WRITING WITH RESPONSIBILITY

The characteristics of good writing discussed in this chapter will be developed in the remaining chapters of this book, where we will look at how to capture an audience's attention, how to write strong openings, and how to write for all forms of the mass media. First, however, we need to consider media writing as a value-laden activity and what that means to you as a responsible writer. In Chapter 2, then, we turn to these and other issues related to writing responsibly.

NOTES

1. Ezra Pound, *The ABC of Reading* (New York: J. Laughlin, 1960), 32.
2. Brit Robson, "Too Young to Live, Too Tough to Die," *Mpls/St. Paul*, Oct. 1987, 67.

3. "Blystone Reports," CNN Presents Best of Blystone 1991, Turner Multimedia, 1992.

4. Ibid.

5. *Seventeen* magazine ad, *Advertising Age*, 9 May 1994, 25.

6. Annie Dillard, *Pilgrim at Tinker Creek* (New York: Harper's Magazine Press, 1974), 5–6.

7. Eugene L. Roberts Jr., "Writing for the Reader," Red Smith Lecture in Journalism, University of Notre Dame, May 1994.

8. Ibid.

9. Theater Listings, *New York*, 30 Nov. 1987, 141.

10. "Goings on About Town," *The New Yorker*, 23 Nov. 1987, 2.

11. http://www.boston.sidewalk.com, accessed Dec. 1997.

12. http://www.seattle.sidewalk.com, accessed Dec. 1997.

13. James Hardie Siding ad, *Coastal Living*, July–August 1998, 4–5.

14. Associated Press, "Colombia Toll Could Hit 20,000, in Volcano-Triggered Mud Slide," *St. Paul Pioneer Press-Dispatch*, 15 Nov. 1985, sec. A1.

15. Bradley Graham, "Colombian Volcano Erupts, Killing Thousands," *The Washington Post*, 15 Nov. 1985. Reprinted in *Best Newspaper Writing, 1986*, ed. Don Fry (St. Petersburg, Fla.: Poynter Institute for Media Studies, 1986), 152.

16. Bradley Graham, "Survivors Recall Night of Horror," *The Washington Post*, 15 Nov. 1985. Reprinted in *Best Newspaper Writing, 1986*, ed. Don Fry (St. Petersburg, Fla.: Poynter Institute for Media Studies, 1986), 157.

17. Francis X. Clines, "Gunman Terrorizes Belfast Crowd Gathered at Funeral," *The New York Times*, 17 March 1988. Reprinted in *Best Newspaper Writing, 1989*, ed. Don Fry (St. Petersburg, Fla.: Poynter Institute for Media Studies, 1989), 66.

18. Ibid., 99.

19. Subaru Outback Limited ad, *Outside Family Vacations*, summer 1998, 10–11.

20. John McPhee, "Giving Good Weight," in *Giving Good Weight* (New York: Farrar, Straus, and Giroux, 1979), 61.

21. Richard Ben Cramer, Reprinted in *Best Newspaper Writing, 1979*, ed. Roy Peter Clark (St. Petersburg, Fla.: Modern Media Institute, 1980), 14–15.

22. Ibid., 40.

23. Osborne Elliott, *The World of Oz* (New York: Viking Press, 1980), 55.

24. *MetroPages*, newsletter of the Metropolitan Council, St. Paul, Minn., summer 1996.

25. Media Advisory, University Relations Office, University of St. Thomas, St. Paul, Minn., 5 Nov. 1996.

26. Donald Murray, *Writing for Your Readers* (Chester, Conn.: Globe-Pequot Press, 1983), 29–33.

27. David Wildermuth, "Rock," KARE-11 News, 25 April 1995.

28. Panasonic ad, "Take It to Extremes," *Rolling Stone*, 27 June 1996, 21.

2
WRITING WITH RESPONSIBILITY

*T*he alarm rings and he stirs. It's 5:45. He could linger under the covers, listening to the radio and a weatherman who predicts rain. People would understand. He knows that.

A surgeon's scar cuts a swath across his lower back. The medicines and painkillers littering his night stand offer help but no cure. The fingers on his right hand are so twisted that he can't tie his shoes.

Some days, he feels like surrendering. But his dead mother's challenge reverberates in his soul. So, too, do the voices of those who believed him stupid or retarded, incapable of being more than a ward of the state. All his life he's struggled to prove them wrong. He will not quit.

And so Bill Porter rises.

— Tom Hallman Jr., "Life of a Salesman"[1]

And so Tom Hallman Jr., a reporter at The Oregonian, begins his award-winning profile of Bill Porter, a 63-year-old door-to-door salesman who trudges miles of Portland's streets every day, selling Watkins spices and soaps. Hallman's profile succeeds in part because it is written in clear and descriptive language and with insightful and straightforward details. But Hallman's profile succeeds on another level as well — an ethical level. Hallman's subject struggles daily with cerebral palsy, yet the writer never falls into the trap of pitying Bill Porter or of portraying him with stereotyped language, avoiding the word *disabled* altogether. Hallman explains why: "We've all seen features about disabled people, but they usually highlight the disability. . . . Instead, I focus on the man. That allowed readers to relate to him in a way they wouldn't have if this had been a story about a man with cerebral palsy. It's a subtle thing, but it's critical."[2]

Indeed, Hallman's focus on the man rather than the physical condition is critical to the success of his profile. It also is an example of the types of ethical choices that all media writers face in producing ethically sensitive media messages.

WRITING FOR THE MEDIA: AN ETHICAL ACTIVITY

News stories, advertisements, press releases and other media formats, no matter how long or short they may be, present their writers with ethical choices. The choices can be large and significant: Should time and space be devoted to a story about one particular salesman? Or the choices can be relatively small: How can the writer craft a lead that will introduce readers to the salesman and convey how he copes with everyday life?

Ethical choices necessarily involve making values judgments. Even media writers who think of themselves as objective reporters of facts or as creators of ads that sell products cannot avoid making values judgments. Media writers bring their own perceptions and experiences to their work, and it is their responsibility to recognize how those perceptions and experiences influence what and how they write about a topic. According to media ethicists Philip Patterson and Lee Wilkins, placing a value on an idea or person "means you are willing to give up other things for it."[3] Many conflicting values may compete in a particular situation. "A forthright articulation of *all* of the values wrapped up in any particular ethical situation will help you see more clearly the choices that you face and the potential compromises you may or may not have to make."[4] Ethical writers, therefore, are honest and thorough in making those values choices.

Consider, for example, the values choices embodied in a recent magazine advertisement for a Saturn automobile, shown in Figure 2-1. The top of the ad features a colorful photo of Saturn customer Patricia Steves holding a bag filled with bananas, eggs, milk and orange juice. Behind her is a silver Saturn sedan, its hood crumpled and its front tires flat. The headline proclaims, "It even saved our groceries," and the main block of text in the ad reads as follows:

> "All right, we did lose one egg," admit both Patricia and Roger Steves. But considering the accident they were involved in, they don't seem to mind. You see, while on the way home from the grocery store one afternoon, they were

*"I*t even saved our groceries.*"*

"All right, we did lose one egg," admit both Patricia and Roger Steves. But considering the accident they were involved in, they don't seem to mind. You see, while on the way home from the grocery store one afternoon, they were rear-ended by a pickup truck and the front of their Saturn was pushed into the car in front of them. Luckily neither Patricia nor her husband was seriously hurt. They did, however, fear the worst for their parcels since the trunk was now "trying very hard to become part of the back seat." Imagine their surprise when they discovered that, with the exception of one unlucky egg, their groceries were as unharmed as they were. THE 1997 SATURN SL1

All Saturns are built around a steel spaceframe. When an accident occurs, the front and rear sections begin to crumple and absorb energy from the impact. This helps the passenger compartment maintain its shape and structural integrity. Which, in turn, helps squash the belief that you have to spend $40,000 to feel safe.

PATRICIA STEVES' 11-EGG FRITTATA RECIPE—Ingredients: 1 cup of diced onion, 1 cup of diced green pepper, 2 cups of diced ham, 11 eggs. Sauté onions and peppers until cooked, mix in ham, divide mixture in half and keep warm. Then beat s of the eggs with a fork, stir in half of the filling mixture and season to taste. Pour into preheated omelet pan. Flip into another pan when bottom of the frittata is set. Cook another 1 or 2 minutes. Repeat with other half of ingredients. Serves 6.

A DIFFERENT KIND *of* COMPANY. A DIFFERENT KIND *of* CAR.
This 1997 Saturn SL1 has an M.S.R.P. of $11,995, including retailer prep and transportation. Of course, the total cost will vary seeing how options are extra, as are things like tax and license. We'd be happy to provide more detail at 1-800-522-5000 or look for us on the Internet at http://www.saturncars.com. ©1996 Saturn Corporation.

Figure 2-1

rear-ended by a pickup truck and the front of their Saturn was pushed into the car in front of them. Luckily neither Patricia nor her husband was seriously hurt. They did, however, fear the worst for their parcels since the trunk was now "trying very hard to become part of the back seat." Imagine their surprise when they discovered that, with the exception of one unlucky egg, their groceries were as unharmed as they were.[5]

The advertising agency team in charge of creating the ad for Saturn made several values choices. First, the ad admits that ordinary people, not just attractive models, drive cars. In doing so, the ad also rejects several strategies typically employed in automobile advertisements, such as that sex promotes car sales and that a glamorous lifestyle can be had by buying a particular model of car. Second, the Saturn ad admits that no car — not even a Saturn — can prevent an accident. But in the boxed copy on the right-hand side of the page, the ad tells us that "Saturns are built around a steel spaceframe. . . . This helps the passenger compartment maintain its shape and integrity" in the event of an accident (see Figure 2-1).

The Saturn ad is thus based on a notion of consumers as rational people who want verifiable information about cars. Unlike the typical car ad that attempts to sway buyers with dazzling photos of cars zipping along beachfront highways, the no-nonsense strategy of the Saturn ad reflects, above all, a respect for its audience. It is an effective ad for an automobile manufacturer that appeals to middle-income buyers; it is also an example of ethical advertising.

In 1996, when Target Stores decided it would stop selling cigarettes, the following company statement was released explaining the decision. The writer expresses several values choices in this simple statement:

> Target is discontinuing the sale of cigarettes in our stores. This is a business decision driven by declining profitability and increased capital requirements. About a quarter of our stores stopped selling cigarettes over the past several years. We expect that cigarettes will be off all our store shelves by the end of September.[6]

The spare, no-nonsense language of the statement signals Target's all-business attitude about the decision. Throughout the statement, the writer maintains the theme that Target's decision is based solely on business factors, not on any objections the company might have about cigarette smoking. In addition, through word choice and an emphasis on the business angle, the writer indicates that Target will not trumpet its decision, despite the ease with which it could endear itself to anti-smoking advocates. (For examples of media writers who have incorporated the anti-smoking angle into their stories, see pp. 54 and 56 in Chapter 3.)

Media messages constantly tell audiences what to think about a range of issues, events and products. Whether the message is about what car to buy or about how a traveling salesman copes with everyday life, media writers have an obligation to examine the ethical issues involved and to make responsible choices about the messages they convey to audiences.

DEALING WITH ETHICAL ISSUES

Deadlines and Ratings

Media writers often work within dramatic boundaries. They get just 30 seconds of air time to convince television viewers to buy a specific brand of tissue, just two hours to crank out a news release telling journalists their company's role in an oil spill or just eight inches of newspaper space to tell readers why a local teacher has won a national award. And their clients, editors and news directors want writing that will hook viewers, sell magazines and move products. But just because media writers are often constrained by deadline and rating pressures does not mean they cannot be ethical.

Sensitive media writers learn how to think and reason quickly when necessary. They learn to rapidly articulate the values involved, discarding the less important ones and focusing on the others. They learn to take account of all the parties involved in a media message and to minimize the harm done to any of those parties by the message. In short, they learn to resolve conflicts among the loyalties, values and responsibilities inherent in media writing.

Some media writers also learn to argue forcefully for the additional time or space they need to present a responsible, accurate message. Tom Hallman Jr., for example, knew he needed time to get to know his subject. Here are three more paragraphs from Hallman's profile of Bill Porter, introduced at the start of the chapter:

> His first stop today, like every day, is a shoeshine stand where employees tie his laces. Twice a week he pays for a shine. At a nearby hotel one of the doormen buttons Porter's top shirt button and slips on his clip-on tie. He

then walks to another bus that drops him off a mile from his territory — a neighborhood near Wilson High School.

He's been up for nearly five hours.

He left home nearly three hours ago.[7]

To get these three detailed paragraphs, Hallman walked Porter's 10-mile route and talked with the people Porter regularly connected with along the way. A week later, Hallman walked the route with Porter, observing his routine. Before Hallman wrote the story, he had interviewed 25 people about Porter. Hallman believed he could not produce an accurate word portrait of Porter without taking the time to understand his subject's life and views. As Hallman explains: "If the Porter story had been on an editor's budget, I could have written a standard Sunday feature. But there was no such pressure because I controlled the timing of the story that I wanted to write by working on other stories."[8] The decision to take the time needed to report and write a thorough story, one far more compelling than a standard Sunday feature, paid off: Hallman says he received at least 2,000 calls and letters from readers, and Porter received another 500.

Advertising writers often enjoy more time than news writers to produce their messages because advertising campaigns are typically group efforts involving elaborate research and planning. Even so, the luxuries of time and planning do not always result in ethically sensitive advertisements. For example, consider a recent advertisement prepared for a major league baseball team, the Minnesota Twins, to promote public financing of a new stadium, which the team said would keep the Twins from moving out of the state. The 15-second television ad features Twins outfielder Marty Cordova's visit with a young cancer patient at a Ronald McDonald House. Before the actual video of the visit begins, an announcer says, "If the Twins leave Minnesota, an 8-year-old from Willmar undergoing chemotherapy will never get a visit from Marty Cordova." Next, Cordova is shown handing a baseball to an unidentified bald boy, after which the announcer comments, "I think Marty brought you something, too." Finally, the ad fades to a black screen, on which the words "Call your legislator" are superimposed.[9]

What values does the ad express? That it is acceptable to exploit a child stricken with cancer? That it is acceptable to use serious illness and charities to foster support for a baseball stadium? Why might the advertising team have thought this strategy was a responsible way to appeal to citizens and state legislators, the target audiences? The answers to these questions lie in how the media writers perceived their loyalties and responsibilities in creating the ad.

Loyalties and Responsibilities

The baseball stadium ad team expressed values choices in writing the ad. The team also demonstrated its loyalty to the Twins, not to the boy, his family or the audience. And the writers decided their responsibility was to create a word-and-picture package that would draw attention to the stadium issue. But the advertising team had not researched or planned its strategy very well before releasing the ad. For one thing, the boy was not from Willmar, Minn., but from another state. For another, the team had not sought the permission of his family or of Ronald McDonald House to use the tape. Worse, the boy had died since his visit with Marty Cordova was videotaped.

In the end, the Minnesota Twins faced a public relations debacle. In addition to the factual errors and insensitivity of the ad, the public perceived it as a sign of the team's disloyalty to the child. Many citizens argued that the Twins' advertising team had a responsibility to treat the child and his family with respect. No amount of compelling copy and poignant videotape could justify using a cancer-stricken child for a fund-raising technique, particularly without the family's permission.[10]

Why does it matter whether as media writers we examine our values choices, loyalties and responsibilities? Ethicists Clifford G. Christians, Mark Fackler and Kim B. Rotzoll say that identifying our loyalties is a way of ensuring clear ethical thought: "To reach a responsible decision, we must clarify which parties will be influenced by our decision and which ones we feel especially obligated to support."[11] They identify five specific parties to whom media writers owe some level of loyalty. First, we have a loyalty to ourselves. Simply put, we should write with such integrity that we can live with ourselves. Second, we have a loyalty to do ethically sound work for our advertising and public relations clients, our news sources, our newspaper and magazine readers, our television and radio audiences, and the businesses and organizations that buy advertising space and time in our media. Third, we owe loyalty to our employers — advertising agencies, newspapers, television stations and so on — making sure we take our employers' interests into account when we make ethical decisions. Fourth, we have a loyalty to our profession and colleagues. After all, the work we do as professionals reflects on others in the profession; every time a news writer takes liberties with the truth, the credibility of all news writers is affected. Finally, we have a loyalty to society, because our messages have tremendous power to reflect and shape society's goals and aims. We should exercise that power with responsibility.[12]

Tom Hallman's profile of Bill Porter (see pp. 29, 33–34) demonstrates loyalty to multiple parties. The profile allows Hallman to respect himself as a creative, careful writer, to respect Porter by portraying him in a thoughtful way, to respect both his newspaper and his colleagues by offering them a compelling story, and to respect society by presenting readers with a thought-provoking examination of a person in their community whose life is typically left uncovered by news organizations.

The Minnesota Twins stadium ad, in contrast, lacks respect for multiple parties. The ad implies that its writers lack thoughtfulness, integrity and respect for the deceased child. How are the client and the public served by this kind of advertising? In this case, the company took a hard hit in terms of public opinion.

Ethical Conflicts

Tom Hallman can balance the values choices, loyalties and responsibilities involved in his story about Bill Porter relatively easily, but for many media writers this can be a difficult task. Values, loyalties and responsibilities may not balance initially, in which case writers have to opt for the best possible sense of balance, albeit an imperfect one. To do so, they need to probe their own motivations and justifications. Media ethicists Patterson and Wilkins suggest that media writers ask themselves three simple questions, based on the work of philosopher Sissela Bok: (1) How does your conscience feel about the "rightness" of your action? (2) Are there alternatives to this action that will not create ethical problems? (3) What are the possible consequences of this action for other people?[13]

The public relations writer of the following passage had to balance multiple values, loyalties and responsibilities. It is a follow-up news release announcing Mall of America's assessment of its newly implemented escort policy. As you read the news release's headline, subheading, lead and subsequent paragraphs, think about how the PR writer successfully balances these issues:

MALL OF AMERICA DECLARES
FIRST WEEKEND OF
ESCORT POLICY A SUCCESS
Slight Modifications to Policy Announced

BLOOMINGTON, Minn. — Mall of America officials say the first weekend under the new Parental Escort Policy resulted in more families in the mall and fewer violations of mall rules.

"We are extremely pleased with the first weekend under the new policy," said Teresa McFarland, Mall of America's public relations manager. "The policy is working exactly as we intended. More families came out to enjoy the mall and it was a safer atmosphere for everyone. This was the first weekend in memory where we didn't have a fight between kids."

The policy requires kids under age 16 to be accompanied by a parent or adult 21 years of age or older after 6 p.m. on Friday and Saturday nights. McFarland said that minor revisions will be made to the Parental Escort Policy to address unusual situations. . . .

The revisions, which go into effect this Friday include:

- To accommodate young parents with children, parents and guardians 16 and over will be allowed to escort children under age 6.
- Disabled individuals over age 16, who need assistance, can be accompanied by an individual under age 16.
- College IDs that include photos and birthdays will be accepted as forms of IDs. Other acceptable IDs include a valid driver's license or a state identification card.
- High school students with reservations for dinner before a high school dance are encouraged to make their reservations before 6 p.m. Escorts to restaurants after 6 p.m. will be provided for any individual who doesn't have proper identification or who is under age 16.[14]

First, the PR writer had to deal with several values choices: What angle of the policy's implementation should be stressed? What information would be most important to the journalists who would read the release? How should mall officials respond to any problems in the policy's implementation? Next, the writer had to deal with issues related to competing loyalties and responsibilities: How could the mall's desire to portray its policy in a positive light be balanced against journalists' desire to learn the solid truth about the success of the policy's implementation? How could the news release accurately portray adult customers' view of the policy while fairly handling the objections of youths under age 16? And how could the obligations of the writer to the client, mall and others be balanced?

By using a headline that pronounces the policy a success and a subheading that mentions the planned modifications, the PR writer immediately balances her loyalty to the client's best interests with her loyalty to journalists, who rely on her for accurate information about the client. The writer's lead stresses the policy's general success, and by selecting a quotation for the second paragraph, the writer emphasizes the mall's motivation for the policy — safety for customers and youths. Subsequent paragraphs build on the subheading. The writer admits that during the first weekend of operation, problems developed that required mall officials to modify the policy. The writer goes on to list each modification, thereby satisfying journalists' need to know how officials responded to the problems. (To see how one TV writer covered the problems, turn to pp. 142 and 144 in Chapter 6.) Notice, too, that the writer sticks to plain language, avoids exaggerating the policy's success and does not overemphasize an apology for the unanticipated problems. Overall, the PR writer does a good job of balancing competing values, loyalties and responsibilities.

OBSERVING THE TASKS OF THE ETHICAL MEDIA WRITER

Ethically sensitive media writers balance several tasks that grow out of their values, loyalties and responsibilities. Three of their most important tasks are to tell the truth, do no harm and promote social justice.

Tell the Truth

Truth is at the philosophical core of journalism, and good news writers strive to tell the truth. After all, journalists enjoy significant freedom of speech because of the First Amendment. An unfettered press free of government interference means that citizens can be told the truth about issues important to maintaining a system of self-government.

Truth is also important to news writers on a practical level in that it ensures their credibility. News writers who are less than truthful, whether deliberately or negligently, lose credibility with their audiences and colleagues. For instance, after WCCO-TV presented a series of reports questioning Northwest Airlines' safety and maintenance procedures in 1996, the station was attacked by the airline and the public for manipulating the facts. Critics said the station's news writers pulled facts about Northwest Airlines from Federal Aviation Administration documents without comparing those facts with the FAA's evaluations of competing airlines. Northwest filed a complaint with the Minnesota News Council, which decided that WCCO had presented a "distorted, untruthful" picture of the airline's safety and maintenance procedures. The News Council's decision was front-page and top-of-the-hour news all over Minneapolis and St. Paul, and ultimately, WCCO's lax treatment of the truth was the focus of a Mike Wallace segment on CBS television's "60 Minutes."[15]

Although advertising and public relations writers may not think of themselves as truth tellers in exactly the same way news writers do, they are still responsible for being truthful. Outstanding advertising and PR writers focus on telling the truth, often from their clients' point of view and with snappy, attention-getting words, but they resist the urge to deceive or manipulate the facts. They know that their ethical obligation to the audience is to present the truth. They also know that their audience will decide whether to patronize an oil company's gas stations based in part on the public messages the company issues about its role in an oil spill, or whether to buy a particular brand of tissue based on the price listed in its advertising. These writers' professional integrity depends on their ability to tell the truth, and they find their credibility enhanced because the journalists with whom they regularly interact come to respect and trust them. Public relations and advertising writers who are not truthful come to be labeled as "hacks" or "flacks."

Recall the straightforward statement issued by Target Stores' public relations department (see p. 32). The PR writer sticks to the facts and resists the temptation of giving Target a grander motive than it had for eliminating cigarettes from its stores. The Mall of America news release also sticks to the facts, yet it presents the employer's

WRITING TIPS

Avoiding Libel

Sensitive media writers avoid recklessly harming people's reputations; that is, they do not commit libel. "Libel is injury to reputation," says the Associated Press. "Words, pictures or cartoons that expose a person to public hatred, shame, disgrace or ridicule, or induce an ill opinion of a person are libelous."[1]

A person can successfully sue for libel by proving five criteria.[2]

1. Publication: The information was printed or broadcast. Publication occurs if only one person other than the injured person reads or sees the damaging language.
2. Identification: The message identified the person, even if he or she was not specifically named. Identification occurs if description points to the injured person. That means writers must carefully verify names, addresses and other identifying details before they use them.
3. Defamation: The information actually was harmful. For example, writing that a person has committed fraud is libelous if the person did not commit it. The inaccuracy thus harms the person's standing in the community.
4. Fault: The writer was negligent or recklessly disregarded the truth. Journalists have some leeway to write about public officials and figures, but they must be able to prove that harmful information is true.
5. Injury: The language injured a reputation. For example, a person's reputation can be injured by a report that claims the person served time in prison when in fact he or she did not.

1. Norm Goldstein, ed., *The Associated Press Stylebook and Libel Manual*, 6th ed. (Reading, Mass.: Addison-Wesley, 1996), 281.
2. Dwight L. Teeter Jr. and Don R. LeDuc, *Law of Mass Media*, 7th ed. (Westbury, N.Y.: Foundation Press, 1992), 105–12.

policy in a positive, realistic light (see p. 37). Finally, the Saturn ad shown in Figure 2-1 features an authentic customer and presents facts about the product. All of these examples may be easily regarded as truthful.

Do No Harm

Because media messages play a crucial role in telling people what to think about, they can wield power in the way people view their neighbors, their community and their elected officials. Media messages also can wield power in the way people view places they have never visited and people they have never met. With this power to alert the public to so much about the world sometimes comes the power to harm.

News writers can do harm when they recklessly intrude on the lives of private people they thrust into the news. As longtime Chicago Tribune columnist Bob Greene tells us, he learned that lesson the hard way early in his career:

> In 1971, the first year I was writing a newspaper column, a fourteen-year-old girl made an appointment to see me. She came to my apartment on a Saturday afternoon.
>
> She was in the ninth grade and had started using drugs two years earlier. Now, she said, she had stopped; her parents had found out and had gone with her to have a talk with the family doctor. But just as she was giving up marijuana, stimulants, and depressants, her classmates were starting. She felt ostracized. She was thinking of taking up the drugs again just so she wouldn't feel left out. . . .
>
> We spoke for several hours, and at the end I said that I did, indeed, want to write a column about her experience. I asked her if she wanted me to use her name. She thought about it for a moment, and then she said she would be more comfortable if I just used her initials. Her initials were J.C.
>
> Two days later the column appeared. The headline was "AT 14, J.C. IS OFF DRUGS." And two days after that, I received a telephone call. It was from the girl's mother.
>
> "I don't know if you'll ever have the experience of having a daughter who is so traumatized that she can't get out of bed," the woman said. "She's been sobbing for twenty-four straight hours."

The girl had gone to school on the morning the column appeared, and immediately discovered that virtually everyone she knew had read it and had easily identified her. They were livid; she had, in effect, told all of their parents about the drugs they were taking. The girl's mother asked me if I understood what this was going to do to the rest of her daughter's high school life, and I said I understood. But I didn't, really; I was brand new at writing about people's personal experiences for large audiences. I had been given a regular column in a major newspaper in one of America's biggest cities, and the column was being distributed to more than one hundred other papers. I was twenty-three years old.[16]

Ten years after he wrote the column about J.C., Greene wrote that he wished he had used "she" to describe the subject in his column. "I do not know what happened to that girl in the months and years after I published her story, but I do know that I caused her pain when I didn't have to, and that the story wasn't worth it."[17] Greene's right: J.C. wasn't a public official, nor was she at the center of a public event. Her initials could have been deleted from the column without sacrificing its power to convey the uphill battle of one teen fighting drugs.

Media writers can also do harm when they make snap judgments or hastily label a person or an issue with a stereotype. Based on generalizations that may be true for some but not *all* people, stereotypes can seem like handy devices to writers. Take a 20ish young woman. Give her double-pierced ears and a "Y" necklace. Clothe her in a dark leather jacket and give her an oversized cup of coffee. Presto: You have a "Generation X-er" for an advertisement. As a media writer, however, you will not sacrifice precious words to establish the character's background; the visual image will do that. Still, it is restrictive to people whose age places them in the Generation X category to portray them consistently as "slackers" who hang out in coffeehouses. Some people in their 20s are farmers, teachers, parents or ministers, and some of them would rather drink milk than coffee. Portrayals that recognize the range of choices people make in their lives are more ethically sensitive than one-dimensional stereotypes that rely on the clichés of vision we discussed in Chapter 1.

While portraying people in their 20s as "Gen-X slackers" may seem only mildly troublesome, stereotypes can have dramatic consequences. Patricia Raybon, a former magazine editor, objects to the stereotypes that news writers use to describe her mostly African American community:

> In my own inner-city neighborhood in Denver — an area that the local press consistently describes as "gang territory" — I have yet to see a recognizable "gang" member or any "gang" activity (drug dealing or drive-by shootings), nor have I been the victim of "gang violence."[18]

The harm done by carelessly labeling a neighborhood "gang territory" is at least threefold: First, the label may not be true. Second, citizens who do not know the area may believe it to be unsafe, and they may avoid and thereby isolate the area. Third, citizens who live in the area are unfairly disparaged and stereotyped. The harm caused by careless labeling is particularly noxious when it is foisted on racial minorities who are often ignored by media messages.

Promote Social Justice

Telling the truth and doing no harm are "negative" ethical duties for media writers, as in "Do *not* lie" and "Do *not* hurt people." In that sense, these are "lowercase-*e*" ethical duties because media writers can perform them without questioning the values of the commercial media system. In that same sense, promoting social justice is a positive ethical duty that requires media writers to question the status quo. It is a "capital-*E*" ethical duty for writers: "*Do* question why the media emphasize the issues they do" and "*do* question how the media can behave in a socially responsible fashion."

Media ethicist Clifford G. Christians contends that because the mass media play a crucial role in interpreting the world, they have a special ethical responsibility to promote social justice for the disadvantaged. Therefore, an ethical use of media messages is to give people and issues that are generally ignored by the mainstream media an opportunity to be heard Drawing on the ideas of political philosopher John Rawls, Christians argues for a "voice for the voiceless" and protection of the least-powerful party. He writes:

> In a day when the powerless have few alternatives left, and virtually no recourse, should the press not serve as a voice, as a megaphone of sorts for those who cry out to be heard?

WRITING TIPS

USING NONRESTRICTIVE LANGUAGE

"Language has great power — but just as it can enlighten, amuse and uplift, so too can it offend, divide and isolate," write Lauren Kessler and Duncan McDonald in "When Words Collide." "As writers — but more, as citizens — it is our responsibility to use language with compassion. Language should help us appreciate differences among people as it promotes equality and reflects a credo of tolerance."[1]

Sensitive media writers avoid five kinds of language that stereotype and exclude groups of people, according to Kessler and McDonald.[2]

1. Sexist language. Instead of writing "Firemen fought the blaze," use "Firefighters fought the blaze."
2. Heterosexist language. Include sexual orientation only when relevant. If you would not write "Jones, a heterosexual, is a lawyer," why write "Jones, a lesbian, is a lawyer"?
3. Racist language. Identify race only when relevant. If you would not write "Jones, who receives welfare, is white," do not write "Jones, who receives welfare, is Hispanic."
4. Ageist language. All older people are not senile or bedridden, and 60-year-olds are not elderly. If you would not write "She's active for a 35-year-old," do not write "She's active for a 75-year-old."
5. "Able-bodied" language. Do not reduce people with physical or mental disabilities to their disabilities. You would not write "The hearing journalist writes well," so do not write "The deaf journalist writes well."

---

1. Lauren Kessler and Duncan McDonald, *When Words Collide: A Media Writer's Guide to Grammar and Style*, 4th ed. (Belmont, Calif.: Wadsworth, 1996), 137–38.
2. Ibid. 138–48.

Shouldn't the communications media be the channel of today's impoverished, so their complaints and pleas for mercy will rise above the noise of a busy and complicated nation?[19]

Of course, media writers who take Christians' duty of social justice seriously will not always find it easy to realize. Many newspapers and TV newscasts steadfastly maintain that they must not make news or appear partial to any party, no matter how rich or poor, powerful or powerless. Magazines and TV programs are full of advertising images of glamorous possessions that many people do not need or cannot afford. Clearly, the positive ethical duty of promoting social justice carries risks for media writers. But the courageous media writer who accepts this ethical duty can demonstrate a loyalty to his or her professional integrity as well as to society.

Consider a piece by Kristin Tillotson, a newspaper feature writer. In mid-January 1997, after the national media had given audiences three weeks' worth of sensational coverage of the apparent murder of 6-year-old beauty-pageant "princess" JonBenet Ramsey, Tillotson wrote a column that demonstrates a sense of social justice. Here is an excerpt:

There were 103 children between the ages of 5 and 8 murdered in the United States in 1995, the last year for which complete statistics are available. In many of those cases, family members were suspects. How many of them rated the national scrutiny that JonBenet Ramsey has? How many appeared on the cover of Newsweek?

None. Why? Their parents hadn't the foresight to schedule a session at Glamour Shots before the tragedy occurred. The primary reason the Ramsey case has warranted daily updates is the videotape of her performing in pageants — imagery that may have seemed innocent while she was alive, but to which her death lent a lurid sexual undertone (or so the excessive coverage implies). Life imitates horror movie: The pretty blonde gets it first, or worst.

Newspapers decry a beauty-obsessed society on the op-ed pages while feeding the fire in the entertainment pages. Television reporters knit their brows and ask viewers, "Are child beauty pageants bad, bad, bad?" then stand aside to run tape of little-girl chorines in high heels playing grown-up sexpots.[20]

Tillotson questions the mainstream media's exploitation of the murdered child, concluding with an impassioned appeal to protect the image of a voiceless victim:

But JonBenet Ramsey was just a kid. She is not a ratings booster, not an excuse for a trumped-up "issues" story, not a fabulous photo opportunity. She is a murdered child.	Shame on everyone who has forgotten, in the name of attracting more consumers, that the killing of a child is an ugly, awful and definitely unglamorous occurrence.[21]

Similarly, Demetria Kalodimos, an anchor at WSMV-TV in Nashville, demonstrated a sense of social justice in a series of stories reported in 1995. Kalodimos had stopped to fill her car's gas tank at a small Tennessee town one day when she noticed several signs in the gas station were written in Spanish. This is how one observer, Valerie Hyman, sums up Kalodimos' series on the issue:

> She thought that was odd, in a nearly all-white county, and started asking questions. She learned the Spanish was to attract the business of Mexican and Central American migrant workers, who gradually were assuming the difficult and sometimes dangerous task of harvesting tobacco. Demetria's subsequent reporting also revealed the workers were living in sub-standard conditions in prosperous agricultural communities. Many dozens of videotapes and notebooks and months later, Demetria's series, "Hard Luck Harvest," aired, right in the middle of the tobacco auction season, and proceeded to win the National Headliner's Award.[22]

Some media writers who undertake the ethical duty of social justice, such as Demetria Kalodimos, win awards because they bring to light important but overlooked community issues. In addition, some organizations that undertake the ethical duty of social justice find their reputations enhanced. The John Hancock Mutual Life Insurance Co., for example, put aside its primary purpose of selling insurance policies to advocate for the Sarajevo Olympic Children's Fund. It ran an ad on the CBS television network during the opening and closing ceremonies of the 1998 Winter Olympic Games in Nagano, Japan. The ad featured emotionally striking images of Sarajevan children playing hopscotch in front of bombed-out apartment buildings and kicking soccer balls around a playing field ringed with the skeletal frames of burned-out bleachers. Over the images were heard the voices of young adults who survived the siege of Sarajevo but remembered their city's shining hour as host of the 1984 Winter Olympics. Said one, "You can destroy many things, you know, but not what happened to us when the Olympics came to Sarajevo. This we have kept alive." The ad concluded with a voice-over that said, "The children of Sarajevo never forgot the Olympics. Please don't forget them. They need hope. They need you."[23]

Spending a few advertising dollars to urge adults to help the thousands of children whose city was devastated by war seems a relatively small, yet important, gesture for a multimillion-dollar corporation to make. As a bonus, an ad that displays a sense of social justice is bound to enhance — not impair — the John Hancock Co.'s reputation, and it did. The week after the Nagano Olympics ended, The New York Times called the ad one of the three best advertisements broadcast during the CBS telecasts.[24]

THE NEXT STEP: WRITING FOR AUDIENCES

As you read further in this book, you will find many examples of well-written media messages, many of which are also sensitive to ethical issues. They take loyalties to multiple parties into account, they tell the truth, they do no harm and, sometimes, they promote social justice. Ethically sensitive writers also take into account their audiences. Writing with an understanding of the audience is an ethical task as well as a practical one. Indeed, a media message that neither insults nor condescends to the audience conveys much about the writer's professionalism and integrity. In Chapter 3, then, we will turn to writing for audiences.

NOTES

1. Tom Hallman Jr., "Life of a Salesman," in *Best Newspaper Writing, 1996,* ed. Christopher Scanlan (St. Petersburg, Fla.: Poynter Institute/Bonus Books, 1996), 171.

2. Tom Hallman Jr., "Lessons Learned," in *Best Newspaper Writing, 1996*, ed. Christopher Scanlan (St. Petersburg, Fla.: Poynter Institute/Bonus Books, 1996), 180.

3. Philip Patterson and Lee Wilkins, *Media Ethics: Issues and Cases*, 2nd ed. (New York: McGraw-Hill, 1998), 99.

4. Ibid.

5. Saturn ad, "It Even Saved Our Groceries," *Metropolitan Home*, Sept.–Oct. 1997, 157.

6. "Target Issues Media Statement on Cigarette Sales," PR Newswire, 28 Aug. 1996.

7. Hallman, "Life of a Salesman," 173–74.

8. Hallman, "Lessons Learned," 180–81.

9. Karl J. Karlson, "Twins Pull TV Ad After Viewers Complain It Exploits Sick Children," *St. Paul Pioneer Press*, 6 Nov. 1997,1A, 8A.

10. Ibid.; Scott Miller, "Player Angry About Role in Pro-Stadium Ad," *St. Paul Pioneer Press*, 7 Nov. 1997, 5A; and Bob Sansevere, "TV Ad Shows Depth of Twins' Pandering," *St. Paul Pioneer Press*, 6 Nov. 1997, 1E, 2E.

11. Clifford G. Christians, Mark Fackler and Kim B. Rotzoll, *Media Ethics: Cases and Moral Reasoning*, 4th ed. (White Plains, N.Y.: Longman, 1995), 20.

12. Ibid., 20–21.

13. Patterson and Wilkins, *Media Ethics*, 3–4.

14. "Mall of America Declares First Weekend of Escort Policy a Success," Mall of America news release, 8 Oct. 1996.

15. Brian Lambert, "Northwest Wins Fight with WCCO," *St. Paul Pioneer Press*, 19 Oct. 1996, 1A, 9A.

16. Bob Greene, "By Any Other Name," *Esquire*, Sept. 1981, 23.

17. Ibid.

18. Patricia Raybon, "A Case of 'Severe Bias,'" *Newsweek*, 2 Oct. 1989, 11.

19. Clifford G. Christians, "Reporting and the Oppressed," in *Responsible Journalism*, ed. Deni Elliott (Beverly Hills, Calif.: Sage Publications, 1986), 111.

20. Kristin Tillotson, "JonBenet's Death an Ugly Example of Beauty Myth," *Star Tribune*, 19 Jan. 1997, F6.

21. Ibid.

22. Valerie Hyman, "Follow Your Curiosity to Find Better Stories," *Workbench*, winter 1998, 8.

23. John Hancock Mutual Life Insurance Co., advertisement for the Sarajevo Olympic Children's Fund, created by Hill, Holliday, Connors, Cosmopulos, Feb. 1998.

24. Stuart Elliott, "TV's Olympic Ads: Straight from Dullsville," *The New York Times*, 24 Feb. 1998, C1, C6; John Hancock Mutual Life Insurance Co.

3

WRITING FOR AUDIENCES

*B*LOOMINGTON, Minn. — *the Bordeaux family had just settled in at one of the Mall of America's immense food courts when the chairs and tables started flying.*

"The crazy things that go through your head," Nancy Bordeaux said. "Cattle stampede. Earthquake. Then it came through that it was a bunch of kids chasing each other."

By the time the brief melee ended that June evening, two of Bordeaux's children had been injured diving for cover and a third had seen a pistol pointed at him — and the nation's premier shopping center was headed for a tough new standard for controlling its younger customers.

"Everybody should feel like they're safe when they're at a mall," said Bordeaux, a tourist from Portland, Ore. "We made a big stink."

The mall responded. Beginning Oct. 4, the Mall of America will institute what amounts to a curfew — posting guards at all 23 entrances to bar anyone younger than 16 from entering after 6 p.m. on Fridays and Saturdays unless they are escorted by someone at least 21 years old.

The policy is risky. Only a few malls have tried such an extreme measure to control teenagers, who also are valued customers. Also, the Mall of America prides itself on its good relations with the community, and although management flatly refuses to say so, most of the young people gathering on weekend nights are black. The opportunities for a misstep or insensitive treatment by security officers are endless.

"It's not a color issue," said mall spokeswoman Teresa McFarland. "It's a behavior issue."

— Karl Vick, "Mall and Order in Minnesota"[1]

At The Westchester Mall in White Plains, N.Y., shoppers pay to park in a garage staffed with gold-jacketed security guards. Inside, marble floors lead to 150 swank stores. Notably absent are boisterous crowds of teenagers.

> *The suburban malls that have long flourished as teenage hangouts are pulling up the welcome mat — but ever so subtly. In prosperous communities, especially, their strategies have less to do with rules and regulations than attitude: Malls are trying to control their customer mix by selecting more exclusive stores that don't draw teens, offering few places for them to congregate and creating opulent interiors meant to evoke a luxury hotel. . . .*
>
> *Certainly, discouraging teens means losing a piece of their estimated $67 billion in spending power. . . . But New York mall consultant Richard Hodos says shoppers' antiteen sentiment has intensified "as people become more aware of gangs and violence."*
>
> *At the giant Mall of America in Bloomington, Minn., bystanders were caught in a teen food-court fight last summer. "There were trays flying," recalls mall spokeswoman Teresa McFarland. The mall, a hangout for as many as 3,000 teens on weekends, has also experienced more serious juvenile incidents, including a shooting over a jacket. It now requires youngsters under 16 to be accompanied by an adult on weekend nights. . . .*
>
> — Louise Lee, "To Keep Teens Away, Malls Turn Snooty"[2]

Both of these stories on Mall of America's parental escort policy appeared in major newspapers in 1996 and offer perspectives on emerging national trends. But even within these parameters, the writers cater to the distinct audience each newspaper serves.

The first story is from The Washington Post, known for its coverage of social issues. Its writer focuses on the national trend of teen misbehavior at malls. The piece leads with an anecdote in an attempt to capture the relevance of this story for a national audience: The Mall of America, a popular vacation destination, draws tourists from across the country with its self-contained aquarium, amusement park, shopping and restaurants. The introductory anecdote thus focuses on the experience of the Bordeaux family from Portland, Ore., visitors to the mall in Minnesota.

The second story, from The Wall Street Journal, also pinpoints the national trend of teen violence at malls. But its writer focuses less on the social and community aspects of the problem and more on its industry outcome: the emergence of upscale, exclusive shopping centers designed to discourage teen-age patrons. This angle is appropriate for readers of The Wall Street Journal, who are interested in retail business trends and financial opportunities. The writer does not lead with the Mall of America policy; instead, she weaves it into the story to satisfy a business-oriented audience.

Like accounting, dentistry or auto repair, writing for the mass media is a service occupation. Media writers provide information or entertainment, or they attempt to influence behavior such as voting, buying or volunteering. A media writer's "customers" are the readers, viewers or listeners who make up the medium's audience. The better a writer understands the audience and incorporates this knowledge into the message, the more likely the audience is to read, watch or listen to the message, and therefore, to understand and act upon it.

In this chapter, we examine the issues you should consider when preparing copy for different media audiences. We point out the primary differences and similarities of audiences for the print, broadcast, online, radio, public relations and advertising media. Much of the work involved in understanding an audience occurs each time a story, press release or advertisement is written. The first step is to read the publication or watch the newscast for which you are writing and try to determine the nature of the audience from the content. Consulting editors and other writers about their experiences with this audience can be helpful as well. Finally, audiences are often analyzed statistically and these data are typically available from newspaper or television circulation and advertising departments, from reference manuals such as Editor and Publisher and Bacon's Directory, and from electronic databases and Internet sites.

WRITING FOR PRINT AUDIENCES

Active Involvement

Even a cursory study of The Wall Street Journal shopping center article would indicate that it is written for an audience involved with issues pertaining to retail business. It is important for media writers to understand the concept of involvement when considering an audience. The word *involvement* comes from the Latin verb *volvere*, which means "to turn, roll or wrap." Metaphorically, then, media writers try to determine what the audience is "wrapped up in." They also try to assess the extent to which the audience is involved in a particular subject, as this will affect both the content and the level of detail included in the writing.

Researchers have consistently found newspaper readers to be highly involved with public affairs. For the media writer, this means a print audience is generally interested and attentive by nature. In contrast to television audiences, which tend to be passive receivers of information,[3] print audiences are active seekers of the news. They exert more effort to read a publication than TV viewers do to watch a television show, and they are willing to expend this energy because of the higher informational payoff they get from the print media.[4] Therefore, media writers can hold the attention of print audiences by emphasizing the importance of the story early in the article, as well as by providing the kind of detail and background not offered by most television and radio newscasts.

A print audience's level of involvement in current events may be consistently high over time or it may be motivated by a particular event in the news.[5] In other words, some people read the newspaper every day, whereas others read it when they want information about a specific event or circumstance. In the latter case, for example, people may first learn of a crisis situation, such as a political assassination or threat of war, through word-of-mouth or a television or radio broadcast, and then turn to a newspaper for more detailed information about the event.[6] These are factors the media writer should take into account when writing for print audiences.

When Washington Post reporter Barton Gellman covered the 1995 assassination of Israeli prime minister Yitzhak Rabin, he knew that many of his readers would have already seen the initial reports on television. Gellman explains why he sought to distinguish his article from the television accounts:

> I know from the Gulf War and other events I've covered that also have had live television saturation coverage that after a while this sort of data stream is so thick and so long and so contradictory. . . . I thought my job was to somehow forge a coherent narrative out of all this by putting together all the best details that I knew to be true.[7]

Here is an excerpt from his story. As you read, notice how Gellman emphasizes the assassination in the lead and incorporates details about the killer and Middle Eastern politics throughout the article.

JERUSALEM, Nov. 4 — A right-wing Jewish extremist shot and killed Prime Minister Yitzhak Rabin tonight as he departed a peace rally attended by more than 100,000 in Tel Aviv, throwing Israel's government and the Middle East peace process into turmoil.

The lone gunman met Rabin, 73, as he walked to his car in the Kings of Israel Square in front of Tel Aviv's city hall. The prime minister had just stepped off a massive sound stage where he had linked hands with fellow ministers to sing "The Song of Peace." Identified by police as Yigal Amir, a 27-year-old law student, the assassin fired three shots into Rabin's back at close range. . . .

Amir, who studied law and computer science at Bar Ilan University, was among the founders of an illegal Jewish settlement called Maale Yisrael that was built this summer in defiance of the then-impending deal to extend Palestinian self-rule to much of the West Bank. Israel Radio and Television reported that he spoke calmly to police tonight, telling them he acted alone, planned the assassination with a sound mind and had no regret.

Under Israeli law, Foreign Minister Shimon Peres becomes acting prime minister but the government is deemed to have fallen. That means it is up to President Ezer Weizman, a political maverick, to select a party leader to attempt to form a new governing coalition. The choice boils down to Peres or Binyamin Netanyahu, leader of the opposition Likud Party. At a 2 a.m. news conference, Weizman declined to discuss his next move. "The man has not yet been buried," he said. Rabin's funeral is scheduled for Monday.

Rabin's death is likely to advance elections that already were shaping up as a decisive test of the government's historic movement toward peace with its Arab neighbors, the Palestine Liberation Organization above all. The Israeli public is profoundly divided, and Rabin's Labor Party-led parliamentary coalition — dependent on opposition defectors and back-room deal-makers loyal only to Rabin — has hung on until now by a thread.[8]

Mass Versus Specialty Publications

Although newspaper audiences differ in their preference for national, local, general and specific types of news, the newspaper is still considered a "mass" medium because of the relative homogeneity of its audience. Other print sources, such as specialty magazines and trade journals, are aimed at more narrowly defined audiences, such as people seeking information on wedding planning, stock car racing or another topic of interest. Writers for specialty publications still need to provide the level of background and detail required by other print audiences, but the content of their writing will reflect their readers' more focused interest.

Consider the following two leads to stories about Target Stores' decision to stop selling cigarettes. This first lead was written for readers of The Arkansas Democrat-Gazette, a general-circulation daily in the Little Rock area:

> After selling cigarettes for 34 years, Target has kicked the habit.
>
> The 714-store national chain has stopped ordering cigarettes and said all packs and cartons will be off its shelves by the end of September.[9]

This second lead was published in Discount Store News, a trade journal for retail industry employees and managers:

> Target cast its recent decision to snuff out smokes as purely a business move, rather than an ethical consideration.
>
> Low profit margins, high theft and expensive-to-implement ordinances to control sales to minors make cigarette retailing less than lucrative.
>
> Indeed, when cigarette makers cut prices in '95, the deflationary move had a major impact on the poor sales results of Sam's Club, since cigarette sales to small businesses are a major part of club revenues.[10]

Each lead takes a different angle on the topic: The Arkansas Democrat-Gazette focuses on the consumer angle, informing its general audience of what Target's move means for purchasing cigarettes, whereas Discount Store News emphasizes the interests of its specialty audience, such as profit margins, sales results and less-than-lucrative prospects for stocking cigarettes. Readers of Discount Store News are likely to be retailers interested in selling cigarettes, while readers of The Democrat-Gazette are likely to be customers interested in buying them. Both stories are written for print audiences and, therefore, provide context and detail. But each writer displays knowledge of the particular audience's interests by modifying the content and emphasis of the article accordingly.

WRITING FOR BROADCAST AUDIENCES

Passive Involvement

As noted earlier, television is considered a more passive medium than print; viewing a TV newscast requires less focused attention than reading a newspaper or magazine. In addition, television frequently competes with several other activities for the audience's attention, and TV viewing is often accompanied by other behaviors such as snacking and socializing. It is not surprising, then, that television viewers tend to be relatively passive seekers of information. Compared to newspaper readers, TV viewers are less involved in current events, and they tend to tune in for entertainment and relaxation, as well as for informational reasons. According to a 1996 report, however, about two-thirds of the American public watch the television news on a regular basis.[11] This figure is generally consistent across age, education and income levels, which means that the television news, like the local daily newspaper, is directed at a mass audience. While a television newscast may not provide the background and depth of a Sunday newspaper article, it has the power to capture people's attention and stimulate interest. Studies have shown that the television news is not all flash and entertainment — that it does contribute to people's overall knowledge of public affairs.[12]

Television can be instrumental in enlightening newcomers, such as children and immigrants, about American politics. As the principal medium through which young people in America and other Western democracies first encounter politics, television also serves as a "bridging medium" for adults who have been socialized in another political system.[13] Television thus introduces people to current events, and when a topic sparks their interest they may turn to newspapers for more information about it, as we saw earlier in our discussion of print audiences.

Unlike the print media, which offer background and detail for information-seeking audiences, television can capture the interest of less-involved audience members, whether they are political new-comers or simply focused on other activities. It does so by way of attention-grabbing tactics, such as by highlighting the sensory quali-ties of a message (music, other sounds, scenery and color)[14] or the emotional response to a message (fear, humor or excitement) in order to attract the viewer's attention to the hard facts of a story. Television story segments also change quickly from one new topic to the next in order to keep viewers interested.

Television news writers often incorporate a human interest angle into their stories. This attracts attention by making the material per-sonally relevant to the viewer. Even in coverage of politics, studies show that people learn about candidates' personal qualities more from televison and about their stands on issues from newspapers.[15]

Here is the script of a local television news report about Target's decision to stop selling cigarettes. Rick Kupchella, the writer and re-porter, incorporates into the story peripheral message qualities, such as scenery and sound, as well as emotion, brevity and human interest.

Throughout the country today — at some 700 locations na-tionwide — Target Stores are pulling cigarettes from the shelves. They are returning them to distributors. It's a matter of policy. Not public policy — just "Target" policy. They are selling cigarettes — no more.

(Interview with Jodi Neuses, Target manager. Neuses says, "We just have to make some choices of which business we want to be in — what is going to be most profitable for us as a company. This obviously is not going to be one at this particu-lar time.")

Target claims its move is strictly a business decision. And cigarettes — for them — are not good business. Theft is rela-tively high — and sales — low. Cigarettes account for only one-half of one percent of sales. What's more, local govern-ments are making more rules — regarding how cigarettes can be sold — and selling to minors can be a real liability.

(Natural sound at store location. Customer says to clerk, "I'd like a carton." Clerk responds, "You want a carton? We're not selling the cartons anymore. This is all we have left.")

There's been no edict as to when they must be gone — but throughout the Twin Cities today the shelves are being emp-tied, and displays carted away.

(Interview at Target with Vanessa Kirk, smoker. Kirk says, "It'll put a little bind on me. I have to go somewhere else and get them now.")

Target insists this is not a moral — or health-conscious call. It's strictly business. To smokers, and manufacturers, it doesn't matter what the motivator is. There's a whole series of outlets now — no longer playing their game.[16]

The script shows that Kupchella has kept his audience in mind. He pins down the point of the story in the first few, short sentences, demonstrating his ability to synthesize and summarize. He uses the natural sounds and scenes of the store, showing a customer asking for a carton of cigarettes. Finally, the reporter incorporates human emotion by capturing the disappointment of shoppers who must go to another store to buy cigarettes. These techniques work together to capture viewers' attention. They help to make the story sharp and colorful and, therefore, to hold the interest of smokers and nonsmokers alike.

Specialty Television Programming

While most television news is directed at a mass audience, some programs appeal to audiences with specific interests. Cable News Network (CNN), for example, has more time to devote to news coverage than the broadcast networks, which must also run entertainment and local programming. CNN can therefore make its programming more specialized, as specialty publications do in the print media. The Target Store cigarette announcement that appeared on CNN's morning financial news segment is almost twice as long as the local news version reported by Kupchella. The CNN story includes the same human interest angle, but it also features an interview with financial analyst Martin Feldman of Smith Barney. Here is the introduction and lead:

Cigarette manufacturers are increasingly finding themselves on the defensive from attacks from both political and financial interests. Target stores will stop selling cigarettes by the end of next month.

Although tobacco stocks are generally rebounding this week, the tobacco industry is still taking it on the chin, and here's the latest hit — Target Stores says it's not going to sell cigarettes anymore because it costs too much to keep the product away from minors. This is a discount chain store, and it wants to get cigarettes off the shelves of its 700 stores by the end of next month.[17]

The CNN story is aimed at people interested in the tobacco industry, some of whom may own stock in cigarette companies. It delivers the news about Target's decision to stop selling cigarettes while also emphasizing the perspective of financial-minded viewers. Most features of television writing apply to both local and cable newscasts, including the use of interviews, scenery and sounds. The contrast lies in the focus of the piece, given the different needs and interests of the audiences.

WRITING FOR ONLINE AUDIENCES

People who surf the Internet and World Wide Web are more akin to involved newsmagazine readers than passive television viewers: They use a specialty medium that requires concentration and attention, and they seek information that interests them. Like the print media, online sources often offer the space necessary to provide in-depth detail about a variety of subjects. In some cases, the online media may have even more space for long articles than print publications, which are constrained by the number of pages allotted for news. The online edition of The Chicago Tribune demonstrates the level of detail that can be provided without space limits. The sports section offers a special site for Bulls fans that features stories about each game of the current season, accounts of citywide victory celebrations and a section on controversial player Dennis Rodman.[18] The news section has longer versions of articles that were published in the printed paper and stories that were cut from the paper because of a lack of space.[19]

The information available on the Internet is updated continuously so that information seekers do not have to wait for the morning paper or the evening news for changes in stock prices or weather predictions. Internet search engines — programs that direct users to specific sites — enable people to pinpoint material that interests them. This makes writing for a Web site similar to writing for a specialty publication: Audience interests drive the content.

WRITING TIPS

USING THE WORLD WIDE WEB

For media writers working on deadline, the World Wide Web can be a quick source of information, if the writer knows where to look. The Web has several reliable sources of so-called expert information on a broad range of subjects. The trick is separating the authorities from the impostors. James Derk, computer editor of the Evansville (Ind.) Courier, says it is "imperative for reporters to proceed cautiously and with a healthy sense of cynicism."[1] Some "expert" Web sites allow anyone who sends in money to be listed in a field of their choosing. Other sites do not check the credentials or references of the people on their lists. It is the media writer's responsibility, then, to verify the qualifications of the expert.

Derk offers the following selection of expert sites:

http://www.FACSNET.org
http://www.profnet.com
http://www.expert-market.com
http://www.yearbooknews.com
http://www.guestfinder.com
http://www.askanexpert.com
http://www.experts.com
http://www.medialink.com
http://www.sources.com
http://www.mediasource.com
http://www.press.org

These sites operate in one of two ways. Some have searchable databases that return a list of possible sources; others send a story idea or theme to thousands of potential sources, who may then contact the inquirer. In addition, a few services hide the searcher's name and theme to prevent competitors from acquiring the information.

1. James Derk, "Charlatan or Authorities? The Web's 'Expert' Directories," *media info.com*, (June 1997), 40.

According to a recent demographic study, most Internet users are men, 31 to 45 years old, with incomes above $50,000 a year. More than half are single, and they spend an average of two to three hours per week online. Since much of the Internet is devoted to commerce, an average of 150 businesses come onto the Internet each day, and 76 percent of people who use the Web say they have purchased a product online.[20]

Despite the availability of news on the Internet, most people learn about current events from the broadcast and print media. One study conducted in 1996 found that less than 9 percent of the American public gets its news from online sources.[21] However, other media experts argue that this figure is climbing. Nielsen Media Research reports that 24 percent of the public had Internet access in 1996, and this percentage has increased by 50 percent each year since then. Yet another study claims the number of Web surfers grows by 10,000 each month.[22]

WRITING FOR RADIO AUDIENCES

The audience for radio has diminished since the introduction of television in the 1950s. Less than half of Americans in a recent survey said they regularly listen to radio news, and 20 percent said they listen to radio talk shows.[23] However, radio is sometimes a widespread source of breaking news, particularly for people who are driving when a tragedy such as a shooting or explosion occurs.[24] Commuters also rely on radio for weather and traffic updates. Like television, radio competes with other activities for a person's attention. Radio use often accompanies primary tasks such as working or driving, and consequently, people tend not to know how much attention they have paid to the radio while they are listening. Therefore, research has shown that people mostly learn about current events from television and newspapers.[25]

Radio writers employ techniques similar to those of television writers to capture a busy audience's attention. They make their stories come alive by using natural sound. Altering background sounds enables a radio story to change "scenery" in the listener's mind and can "transport" people to distant places. Radio writers also vary sound by interviewing people, while at the same time capturing a human interest angle.

Like their counterparts in television, radio writers attract their audience's attention by synthesizing complex topics into brief, sharp sentences. Here is an example from National Public Radio, which aired the Target Stores cigarette story on All Things Considered:

> Anti-smoking activists praise the decision by Minneapolis Target store chain to stop selling cigarettes. Target will discontinue cigarette sales in stores by the end of September, citing declining profits. Other major discount stores probably won't follow the lead.
>
> (Carolyn Bookter, Target spokesperson, says in sound bite, "The decision was driven by declining profits. We were spending more to sell the cigarettes. Local sales laws cost money to the company.")
>
> Analysts speculate on the reasons for Target dropping cigarettes.
>
> (Walter Lowe, of Retail Industry newsletter, says, "They responded to general pressure.")[26]

The reporter, Jim Zarolli, continues with an interview from Martin Feldman, the financial analyst who also appeared on CNN. The short sentences and the natural sound from three different interviews keep the story sharp and lively.

WRITING FOR PUBLIC RELATIONS AUDIENCES

Unlike audiences for the print and electronic media, which differ according to the content of the publication, newscast or Web site and may have general or specialized interests, the audiences for public relations writing are distinguished by their relationship with specific organizations. The purpose of public relations is to build a relationship, using communication, between an organization and its various constituencies. An organization may be a college, a corporation or a neighborhood group. Its constituencies are sometimes called *target audiences* or *target publics*.

Organizations have both *internal* and *external* audiences. The internal audiences of a college, for example, would include students, staff and faculty. The external audiences would include the media, alumni,

WRITING TIPS

MEETING DEADLINES

Media writers face many pressures on the job, including the pressure of meeting deadlines. Their writing must be clear, concise, well-organized and completed on time. Media writers usually do not compose at a leisurely pace; a client may call at 9 a.m. and need a news release and fact sheet by noon. A fire may ignite at a historic building at 10:30 p.m., killing several people. After gathering the information, the reporter may have only 30 minutes to write the story.

These demands are often stressful for beginning media writers, but writing well and writing fast become easier over time and with practice. Your confidence will grow each time you complete a story on a tight deadline, and your fear and procrastination will fade.

Public relations writers who work with the media must meet the deadlines established by their companies as well as the deadlines of newspapers and radio and television stations. If PR writers want the media to cover news generated by their companies, they must know and follow media deadlines. Few newspaper reporters who work for morning papers would appreciate a call from a public relations person at 4:30 p.m. — the usual time the reporter is trying to finish stories for the next day's paper. A public relations writer will get a better reception from the reporter at 10 a.m., when the news day is just beginning. Reporters for a 6 p.m. television news broadcast more willingly take calls earlier in the day — say, from 10 a.m. to 1 p.m. — than later in the afternoon as their deadline approaches.

donors, neighbors and government leaders. Public relations professionals write materials for these different audiences, not only to advance the immediate objectives of the organization, but also to build long-term relationships with constituencies. A college may publish a weekly newspaper to keep students, staff and faculty informed of upcoming events. By doing so, it also builds community involvement among these groups, which is a long-term goal of the college.

External Audiences

Let us look again at the Mall of America parental escort policy, which you read about earlier in the chapter-opening quotations. Here, however, our focus is not how reporters covered the story, but how the mall's public relations writers prepared materials for several external and internal audiences. Therefore, in this case, the reporters who wrote the two stories at the start of this chapter were among the Mall of America's external audiences, which also included other newspapers (local and national), magazines and television and radio stations across the country. The mass media, then, were the mall's major external audience, and in September 1996, the mall sent a news release to hundreds of media outlets to announce the new policy. Here is the opening of that release:

For immediate release:
Contact: Teresa McFarland
Wednesday, September 4, 1996

MALL OF AMERICA TO IMPLEMENT
PARENTAL ESCORT POLICY

BLOOMINGTON, Minn. — Mall of America will implement a Parental Escort Policy effective October 4, 1996 to reduce the growing number of unsupervised youth at the Mall on weekend nights.

Under the new policy, youth under 16 will need to be accompanied to the mall by a parent or guardian 21 years or older, from 6 p.m. until closing time on Friday and Saturday nights. Youth under 16 who do not have a parent or guardian with them will not be allowed to enter or remain at the Mall of America after 6 p.m.

"We welcome kids to Mall of America and recognize that the mall is an exciting place for them to visit, yet over the years we have seen a steady increase in the number of unsupervised kids coming to Mall of America on weekend evenings," said Maureen Bausch, associate general manager.[27]

Newsworthy Information

In most cases, the primary audience for a news release includes news reporters, editors and news directors because the objective is to get the release placed in a publication or newscast. The public relations writer thus seeks to interest editors and reporters in the story. The best way to do this is to focus on the news and to write it like a news story. Notice that the mall's news release does not begin like this: "Mall of America is pleased to announce its new parental escort policy, an innovative plan developed by Joan Smith, mall supervisor." Instead, it pinpoints the news: *what* the policy is, *when* it will become effective and *why* it is being implemented. Editors and reporters are likely to respect its clear writing and to continue reading; they may also be persuaded to include the information in a news story. News releases are seldom quoted verbatim in a newspaper or a television newscast. Their function is to provide story *ideas* for the media; reporters then gather additional information and craft their own stories. Releases with newsworthy ideas gain credibility with reporters and are most likely to be used in news stories. Mall of America's complete news release includes substantial detail about the parental escort policy. This is important to reporters and editors covering the story for readers who want detailed information.

However, public relations writers have to meet the needs of other audiences in addition to the news media. They also seldom work alone. The news releases they write are usually revised by co-workers or supervisors, whether the writers work directly for a company or for a public relations firm that represents a company. In the latter case especially, a key audience for the public relations writer is the client — usually someone who represents the organization for which the release is being written. The client reads the draft release to verify the accuracy of the information and to ensure the wording reflects the organization's interests. This is often called the *approval process*.

Clear, Concise Information

In addition to a news release, the Mall of America PR writers prepared a fact sheet about the policy. Fact sheets are particularly useful to television editors, allowing them to see at a glance whether the news can be briefly summarized. A media advisory, or media alert, is similar to a fact sheet: Usually one page in length, it emphasizes the main points of the event or news. Fact sheets and advisories also specify the visual images connected with the news; this reflects the PR writer's awareness of television as a visual medium. When a fact sheet indicates the existence of action connected with the news, a television editor is more likely to assign a reporter and camera crew to cover the story.

Here is the Mall of America's fact sheet.[28] As you read, notice the use of headings and bulleted lists to summarize and organize the information.

MALL OF AMERICA
Parental Escort Policy Fact Sheet

Implementation
- Mall of America will implement a Parental Escort Policy beginning October 4, 1996 to ensure that all guests have a safe and enjoyable experience at the mall.
- The policy will be in effect on Friday and Saturday evenings from 6 p.m. until the mall closes.
- 30 new security officers have been hired to assist with the Parental Escort Policy.

The Rules
- Youth under age 16 must be accompanied by an adult 21 years or older after 6 p.m. on Friday and Saturday evenings.
- Youth 16 and over must have a valid driver's license, legally produced picture ID or a Mall of America employee card.

The Youth Communication Center
- Youth asked to leave Mall of America will be allowed to use the Youth Communication Center, located on the lower level of the mall near management offices, to call for a ride home.
- The Youth Communication Center will be outfitted with 10 phones and staffed with mall managers, youth liaison officers and "Mighty Moms."

The Message Center
- Youth asked to leave Mall of America may leave a message for separated parties at a Message Center Database.
- Message Center Databases are located at each of the four Guest Service stations at each end of the mall.

Groups
- Groups that include youth under the age of 16 must be accompanied by an adult 21 years or older.
- The adult must be able to exercise supervision and control over the group or they will be escorted from the mall.
- Supervision and control issues and questions will be determined by trained security staff and mall managers on duty.

The fact sheet provides editors and reporters with a succinct synopsis of the policy: Once it is enforced, security officers will patrol the mall, teen-agers will be asked for identification, and those who are asked to leave will be allowed to call for a ride home.

In addition to communicating with the media through news releases and fact sheets, public relations professionals often must answer the media's questions as spokespeople for an organization. They draft question-and-answer sheets to prepare them for these interviews, anticipating the questions that reporters may ask and devising answers that reflect the organization's position. Here are a few of the questions and answers prepared by the Mall of America writers:

QUESTIONS AND ANSWERS
PARENTAL ESCORT POLICY

Why Is Mall of America Implementing This Policy?
Mall of America is implementing the Parental Escort Policy to ensure that our guests have a safe and enjoyable experience while they are here. The reality is that right now we have too many unsupervised kids coming to the mall on weekend evenings which is increasing the potential for a security incident. We believe by encouraging families to come out together, we can prevent many problems before they happen.

Why Did You Decide on Youth Under Age 16?
We need to reduce the number of unsupervised youth at the mall. We decided it was logical to start with youth under age 16. Kids under age 16 can't drive by themselves to the mall. If they are going to be here on a weekend evening, we think it makes sense that they should have adult supervision.

Why Wait Until September, Why Not Start It Right Now?
As you can imagine, a lot of preparation is needed to properly implement a policy like this. We've been seriously looking at this issue for nearly a year and we want to make sure we have all the procedures in place before we implement the policy. We also want to make sure the public has adequate time to learn about the policy so we are not turning people away.[28]

By preparing these and other questions and answers in advance, the PR professionals will be better prepared to give reporters and editors the information they need. Advance planning is necessary to allow enough time for the spokesperson to draft succinct, thoughtful answers and to have them approved by management if necessary. Preparation also means that reporters get as much information as the organization can provide, which, in turn, enhances the long-term relationship between the organization and the media.

Customers and Community Groups

In addition to the media, Mall of America had to address its other external audiences, such as customers and community youth organizations. The mall used the publicity generated by media coverage as well as advertisements to reach its customers and sent letters to community youth organizations to address their concerns. In preparing the text for the letter, which is reprinted here, the PR writers sought to inform youth leaders that the parental escort policy was designed to curb dangerous behavior, not to banish teens from Mall of America:

August 20, 1996

Dear Community Youth Leader:

In the last few years, Mall of America has been increasingly challenged by the growing number of unsupervised youth in the mall on Friday and Saturday evenings.

While many of the kids who come to the mall behave and respect the rules, there is a percentage within this group that is loud, abusive, harassing and intimidating. Mall of America's number one priority is to provide our guests with a safe and fun place to visit. We truly feel the large number of youth without parental supervision is jeopardizing that goal. . . .

Your daily outreach to youth in the community makes you an important resource for helping us effectively communicate the new policy. We welcome your thoughts regarding the policy and its affect on the youth in our community. Please feel free to contact me at 555-xxxx.

Sincerely,

Randy Barber, Community Outreach Manager[29]

Public relations writers address many audiences and write in various formats. In the case of the Mall of America, the materials written for community leaders differ significantly in content and form from those prepared for the media. Communication with community leaders requires the more personal format of a letter, rather than the more impersonal flyer or announcement. In addition to sending letters, the mall's public relations staff held group meetings with community leaders to discuss the policy and answer their questions.

Internal Audiences

The internal audiences for the Mall of America included store employees and managers. The mall did not rely on the media to inform these audiences of the new parental escort policy. Rather, it addressed employees' concerns in a newsletter and provided managers and merchants with a packet of informational materials. Each merchant also received a letter and was afforded the opportunity to give feedback via a contact phone number. A separate employee survey on the effects of the new policy gave this audience a chance to voice its views. Together, the letters, survey and informational materials promoted two-way communication between management and employees, a necessary component in building trust and consensus among groups.[30] The exchange of ideas often results in better policies and leads to solid long-term relationships between an organization and its audiences.

WRITING FOR ADVERTISING AUDIENCES

Advertising audiences are defined in terms of their relationship to the product, service or issue being promoted in an advertisement. Designing an effective advertisement, therefore, depends on a thorough knowledge of the target audience. Since the purpose of any ad is to appeal to the needs or interests of a specific group of people, writers cannot craft persuasive copy without first identifying the characteristics of the audience.

Some advertisements are directed at a mass or general audience. In the United States, where television is the most frequently used medium, it is not surprising that such products as food and automobiles are commonly advertised on television. However, the commercials for specific automobiles may vary depending on the audience for a particular show. Baseball viewers are likely to see ads for pickup trucks, whereas people watching an evening hospital drama may see commercials for luxury cars. Similarly, specialty publications often feature advertisements with a narrow appeal. In Modern Bride, readers find ads for wedding dresses, china patterns and honeymoon vacations; in Rolling Stone, they see ads for car stereos, designer jeans and imported beer.

Persuasive Communication

Advertising can influence the audience's thoughts, feelings and behavior. Many theories exist to help explain this process. They examine why people pay attention to a message, understand a message and remember the information in a message. Researchers at Yale University in the 1950s were among the first to study the effects of advertising on audiences.[31] They examined how the relationships among four specific factors might lead to persuasive communication: recipient, channel, message and source. Many later theories have been based on the roles these factors play in advertising. For the beginning advertising writer, they provide a framework for analyzing and constructing persuasive messages.

The *recipient* is the audience member; it is also the first factor the advertising writer should consider in designing an ad. Knowing about the recipient enables the writer to choose the best *channel*, such as a print ad, a broadcast ad, a billboard or an Internet site. Next, the writer drafts a *message* designed especially for the recipient and channel. Finally, the information in the message is derived from a *source* that is likely to influence the recipient. The source might be a person (such as a celebrity), a group (the Peanuts gang) or an organization (a school). In a pet food commercial, the source could be a cat or a dog.

Advertising copywriters, like public relations writers, seldom work alone. Their audiences include co-workers, managers and clients, who may edit, revise and approve the copy, as well as the target audience. Like all media writers, advertising writers need to learn about their audiences. The target audience is critically important in advertising because the effectiveness of any ad depends on reaching the right people. With the dramatic growth in cable, television and specialty publications, all media now make extensive use of research to draw their segment of the public, but in advertising, audience research has long been a major component of the overall strategy employed in creating an ad.

Most audience research is designed to collect four types of information about a target audience: (1) demographic information, (2) psychographic information, (3) media use habits, and (4) level of involvement with the product, service or issue being advertised. Demographic data describe people in terms of their age, gender, education, income, occupation, residence and marital status, while psychographic data identify their lifestyles, attitudes and values. Data on media use habits give an idea of the publications people read and the television shows they watch. Measures of an audience's involvement assess both level of interest in and degree of personal relevance of the product, service or issue. Personal relevance describes how connected people feel to the subject being advertised. Highly involved audiences tend to be influenced by advertisements rich in information about a topic, whereas less involved audiences are influenced by advertisements that feature qualities "peripheral" to the central message.[32] Peripheral advertisements focus on music, scenery, color, graphics or the status of the spokesperson, such as a celebrity or sports star. Like television programming directed at an uninvolved audience, peripheral advertising is designed to capture the audience's attention. It can also create positive feelings while enhancing the image of the product, service or issue being advertised. However, when the subject of an advertisement has a high degree of personal relevance to an audience, a print ad may be most effective. Advertisements in newspapers or magazines, for example, can include more words and complex information than radio or television spots. Broadcast ads, billboards and pictures in high-gloss magazines operate well when the audience is less involved. These visual media can incorporate more sophisticated peripheral qualities, such as scenery, graphics, color and sometimes sound.

Analyzing the Advertisement

After the data on an audience's demographics, psychographics, media use habits and level of involvement are collected, they are used to help determine the best channel, message and source for the ad. Mall of America's advertisement, developed by Olson and Co. in Minneapolis (see Fig. 3-1), will help us analyze this process. The demographic data revealed that the recipients of the ad would be primarily parents and young people under the age of 16 living in the Twin Cities metropolitan area. Other demographic data indicated that behavior problems among children of tourist families were uncommon and that teen-agers visiting the mall came from homes that varied in income and education levels. The psychographic profile of the target

Figure 3-1

audience described teen-agers who are independent, who like to socialize, who may be dealing with peer pressures, and who like to gather indoors, especially during the winter months. In terms of media use habits, teen-agers were described as relying on radio and television, whereas their parents used radio, television and newspapers.

Because this was a new policy, involvement in the issue was not especially high at the outset of the publicity campaign. The mall needed to attract attention to the policy, which would have personal relevance for both teen-agers and their parents. Once the advertisement captured people's attention, the mall would need to provide extensive information about the policy.

The advertisement that ran in the Minneapolis and St. Paul daily newspapers in the fall of 1996 is shown in Figure 3-1. Olson and Co. also developed a radio spot that aired on stations targeting both teen-agers and parents (see p. 143 in Chapter 6) and a print ad for the sides of city buses. Selected on the basis of the audience's characteristics, the buses were an excellent channel because teen-agers often rode a bus to the mall and tended to congregate near the mall's bus stop.

As you can see in the figure, the message and content of the advertisement include both peripheral and informational elements. It looks like a poster for a motion picture thriller, capturing the audience's attention through humor, graphics and a lighthearted tone. It also addresses the independent nature of many teen-agers who might be embarrassed to be seen accompanying their parents to the mall. At the bottom of the ad, information about the escort policy and upcoming activities of interest to teen-agers and their families is provided.

The primary source of information for the print and radio ads was teen-age boys. The radio version featured a conversation between two young men about the escort policy. The mall intended to keep the radio message upbeat, and using humorous dialogue between the teen-agers accomplished that goal. The sources helped to make the ad personally relevant not only to young people, but also to parents familiar with the needs of their children. The advertisement would have had an entirely different tone if the source had been a security guard, mall manager or parent, for example.

The print advertisement was included in a package of information given to reporters before the escort policy took effect, and some newspapers ran the ad as a graphic element alongside the news story. "The reporters were laughing," said Teresa McFarland, mall spokesperson, in an interview about the ad. "They're the hardest audience to reach.

When you can get them to laugh at an advertising campaign, you know you're doing all right."[33]

Overall, the print and radio advertisements served the goals of the Mall of America: to create awareness about the new parental escort policy and to persuade teen-agers to bring an adult with them on weekend nights. The ad works because it persuasively combines the factors of recipient, channel, message and source. It is related to its target audiences and designed to run in the right channels; also, its message attracts attention through humor and graphics and conveys information about the escort policy through an effective use of sources and tone.

THE NEXT STEP: WRITING AND GATHERING INFORMATION

Across all media formats, the better a writer knows the audience, the more effective the writer's message will be. One of the first tasks of writing, therefore, is to gather information about the people who will read, watch or listen to your story. But the information-gathering process involves much more than audience analysis. Media writers must gather material for their ad copy, news stories and press releases from a variety of sources, including personal interviews, government archives and electronic databases. Good writers know that well-researched articles tend to "write themselves"; that is, the better the information gathered, the easier it is to write a good story. In Chapter 4, then, we examine ways to gather and write about information.

NOTES

1. Karl Vick, "Mall and Order in Minnesota," *The Washington Post*, 18 Sept. 1996, A1.
2. Louise Lee, "To Keep Teens Away, Malls Turn Snooty," *The Wall Street Journal*, 17 Oct. 1996, B1.
3. Steven Chaffee and Stacey Frank, "How Americans Get Political Information: Print Versus Broadcast News," *The Annals of the American Academy of Political and Social Science*, 546 (1996): 48–58.
4. Stacey Frank Kanihan and Steven H. Chaffee, "Situational Influence of Political Involvement on Information Seeking: A Field Experiment" (paper deliv-

ered at the annual meeting of the Association for Education in Journalism and Mass Communication, Anaheim, Calif., Aug. 1996), 24–29.

5. Ibid.

6. Bradley S. Greenberg, "Diffusion of the News About the Kennedy Assassination," *Public Opinion Quarterly*, 28 (1964): 225–32; Karl-Erik Rosengren, "The Comparative Study of News Diffusion," *European Journal of Communication*, 2 (1987): 227–55; Walter Gantz, "The Diffusion of News About the Attempted Reagan Assassination," *Journal of Communication*, 33 (1983): 55–66.

7. Barton Gellman, quoted in *Best Newspaper Writing, 1996*, ed. Christopher Scanlan (St. Petersburg, Fla.: Poynter Institute/Bonus Books, 1996), 20.

8. Barton Gellman, "Israeli Prime Minister Yitzhak Rabin Is Killed," *The Washington Post*, 5 Nov. 1995. Reprinted in *Best Newspaper Writing, 1996*, ed. Christopher Scanlan (St. Petersburg, Fla.: Poynter Institute/Bonus Books, 1996), 3–4.

9. Aron Kahn, "Target to End Cigarette Sales in All 714 Stores," *The Arkansas Democrat-Gazette*, 29 Aug. 1996, 1D.

10. Richard Halverson, "Wealth, Not Health, Motivates Target to Extinguish Cigarettes," *Discount Store News*, 16 Sept. 1996, 6.

11. Guido H. Stempel III and Thomas Hargrove, "Mass Media Audiences in a Changing Media Environment," *Journalism and Mass Communication Quarterly*, 73 (1996): 549–58.

12. Chaffee and Frank, "How Americans Get Political Information," 58.

13. Ibid., 57.

14. Richard E. Petty and John T. Cacioppo, *Attitudes and Persuasion: Classic and Contemporary Approaches* (Dubuque, Iowa: Wm. C. Brown, 1981).

15. Chaffee and Frank, "How Americans Get Political Information," 55.

16. Rick Kupchella, "No Smoke," KARE 6 p.m. news, 28 Aug. 1996.

17. Stuart Varney, "Tobacco Industry Takes Hit from Target Stores," CNN 6:16 a.m. news, 29 Aug. 1996.

18. Christopher Harper, "Doing It All," *American Journalism Review*, Dec. 1996, 24–29.

19. Ibid., 29.

20. "Internet Demographics," estimates compiled by research firm Hambreccht and Quist (online database), accessed 27 Jan. 1997 at http://www.fishman.com; and "The Consumer Technology Index," from Computer Intelligence InfoCorp (online database) accessed 27 Jan. 1997 at http://www.harrison digital.com.

21. Stempel and Hargrove, "Mass Media Audiences," 552.

22. "Internet Hyper-Growth Continues, Fueled by New, Mainstream Users," Nielsen Media Research and CommerceNet, Aug. 1996 (online database) accessed 27 Jan. 1997 at http://www.nielsenmedia.com; Hambreccht and Quist, "Internet Demographics."

23. Stempel and Hargrove, "Mass Media Audiences," 552–54.

24. Gantz, "The Diffusion of News," 55–66.

25. Chaffee and Frank, "How Americans Get Political Information," 54.

26. Jim Zarolli, "Target to Stop Selling Cigarettes," All Things Considered, National Public Radio, 29 Aug. 1996.

27. "Mall of America to Implement Parental Escort Policy," news release, Mall of America, 4 Sept. 1996, 1.
28. "Mall of America Parental Escort Policy Fact Sheet," Mall of America, 4 Sept. 1996.
29. Randy Barber, Mall of America letter to community organizations, 20 Aug. 1996.
30. James E. Grunig and Larissa A. Grunig, "Models of Public Relations and Communication," in *Excellence in Public Relations and Communication Management*, ed. James E. Grunig (Hillsdale, N.J.: Erlbaum, 1992), 285–325.
31. Carl I. Hovland, I.L. Janis and J.J. Kelley, *Communication and Persuasion* (New Haven, Conn.: Yale University Press, 1953).
32. Petty and Cacioppo, *Attitudes.*
33. Teresa McFarland, interview by Stacey Frank Kanihan, 17 Jan. 1997.

4

WRITING AND GATHERING INFORMATION

*M**y first interview was with the Man, the King of The Blues, B.B.
King himself. As I made my way across the cavernous lobby of the
Las Vegas Hilton, I tried to picture what our meeting would be like —
probably in a tiny sanctum with no distractions. I felt prepared, but boy,
was I nervous. After the guard waved me through, I wiped the sweat
from my nose, tried to calm my jitters, and then stepped into a pulsating
back-stage suite larger than some nightclubs. Two dozen people were sip-
ping drinks and chatting, most of them B.B.'s family members or old
friends from Mississippi. They were all dressed better than I was and
seemed utterly relaxed. I stuck out like Dolly Parton in an Amish
church.*

*Suddenly woozy, I pictured myself tagging along behind my inter-
viewee and pestering him as he tried to attend to his duties as host and
star attraction. B.B. surprised me by hushing the crowd and announcing
the interview, adding, "Maybe you'll learn something about me you
didn't know before." He had ordered sandwiches and more drinks and
arranged for folding chairs. If I hadn't felt self-conscious enough already, I now found myself at the center of a catered event, conducting my
first interview before seated, attentive observers who knew my subject in
ways I never could.*

*At one point B.B. said, "My people were ashamed of the blues." I
kept smiling and nodding like I had an inkling of what he was talking
about. African-Americans ashamed of the blues? Wait a minute. They
invented the blues. Pretending to listen, I started perspiring and won-
dered whether to stop the conversation and confess that I didn't get it. I
thought, what's the worst that can happen? Well, I answered, I can
make a fool of myself and prove to the multitude that I am a clueless
geek who doesn't belong here.*

*But I took the plunge and admitted that he had lost me, and the
King of The Blues, as gracious as he is gifted, explained that in his*

youth, blacks were made to feel inferior by a complex web of stigmas and stresses. If performers sounded "white" — like Nat King Cole, say — that was good. But if they sounded black, well, that was something to sweep under the rug. And the blues was the blackest music anyone had ever heard. As a young man, he had played the blues, all right, but he did it "behind closed doors, after twelve." I set my prepared questions on the floor and steered the conversation toward a topic I hadn't planned to cover, the relationship between race and music. It was the most fascinating part of any interview I've ever done.

— Tom Wheeler, "When in Doubt, Ask!"[1]

Writer Tom Wheeler learns in his interview with B.B. King that "getting the goods is more important than looking cool,"[2] and media writers can learn from Wheeler's example: Savvy media writers prepare extensively before they conduct interviews, and during interviews, they are observant and flexible enough to pursue startling new angles. In the end, the writers' preparation, observation and flexibility result in "getting the goods" — the detailed information crucial to good media writing.

TAILORING THE INFORMATION SEARCH

Before conducting an interview or surfing the Internet for sources, media writers should ask themselves four questions to help focus their search for information. Shirley Biagi offers the following advice in her book "Interviews That Work."[3]

First, the writer asks a practical question: *How much time do I have to track down information?* A TV reporter whose news director has just bellowed out an order to get over to the city's airport and interview an arriving diplomat will not have time to conduct an extended information search. If the reporter is a well-read follower of current events, he or she will be able to think of questions on the spur of the moment. However, a public relations writer assigned to work on his or her company's annual report has several weeks to track down information that will help him or her understand the ins and outs of the company's financial position. To write an annual report, he or she must conduct an extended search for information.

Second, the writer asks: *What do I need to know?* An advertising copywriter who needs the phone number of a discount store to insert in a newspaper advertisement has no reason to waste time on a documentary search: He should get a telephone book. But a university public relations writer preparing a fact sheet for AIDS Awareness Week on campus may need to know how many Americans are living with HIV; she should seek documentary sources, such as the Statistical Abstract of the United States and the Web site of the Centers for Disease Control.

Third, the writer asks: *How will I use this information?* If the information is critical to the news release, the writer can launch an extensive search for the statistics on HIV patients. But if the focus is on campus plans for making students aware of AIDS, the writer can make a quick sweep of sources and waste little time trolling for tangential information.

Finally, the writer asks: *Who is the audience for this information?* If the information is to be used to form questions for an interview with a scientific researcher whose work is developing new HIV-fighting drugs, the writer should spend several hours learning about the properties of the drugs and becoming familiar with the relevant scientific terms. But if the information is to be used to provide a supporting paragraph in a fact sheet, the writer should spend time looking for translations of scientific jargon into everyday language for the journalists who will receive the AIDS Awareness Week media kit.

USING A VARIETY OF INFORMATION SOURCES

Decisions as to how best to gather information will depend on the writer, the subject and the time available. But all media writers need solid information before they can begin writing. Advertising legend Leo Burnett "has said that in addition to a facility for putting words together, 'organizing facts and finding things that were interesting to people' contributed to his success."[4]

Knowledgeable media writers find facts and information in a variety of documentary sources. A common source of public documents and other information is the government. But numerous other places to look include the local library, the Internet, businesses, nonprofit and other media publications and broadcasts. When a media writer wants to know about a person, the list of possible sources is lengthy and always changing. Think about yourself as a student and a citizen, and the number of cards you carry around in your wallet or purse: your driver's license, college ID card, voter registration card, Social Security card, club membership card and perhaps, a visa or passport. Many of these cards are connected to public documents that are available to anyone wanting to look. Depending on your state of residence, for example, driver's license records are fairly easily obtained, complete with name, age, address, physical description and driving record. A reporter investigating a traffic accident might find this information useful.

In their book "Search Strategies in Mass Communication," authors Jean Ward and Kathleen Hansen identify what they call "potential contributors" to a mass media message.[5] Ward and Hansen divide these contributors or sources of information into three categories. *Informal sources* include observation, casual reading, and networks of friends and co-workers. *Institutional sources* are the traditional tools of the journalist's trade: government documents, colleges, interest groups and businesses. *Information and data tools* are the various libraries, commercial databases, Internet resources and data files available through a computer-generated search.

A listing of all the possible places to locate information would be endless, but we can identify the types of sources that tend to be most useful to media writers. These sources include commercial databases, Internet sources, government documents, interviews and observation.

Commercial Database Services

Most writers begin researching a subject by finding out what others have written about the topic. The quickest and easiest way to locate such information is through a commercial database service. Databases have been around in computer form since the late 1970s, and in printed versions for much longer. "The Readers' Guide to Periodical Literature," for example, is a well-known source of information published on general topics.

Nora Paul, library director at the Poynter Institute for Media Studies in St. Petersburg, Fla., describes commercial database services as "information malls."[6] Each service offers a specific kind of information, such as bibliographies, articles, books, government documents, photographs or telephone numbers. An advertising copywriter about to embark on a campaign for a new cancer drug could look for information in the Cancerlit database, which contains citations to cancer studies, articles and treatment reports dating back to 1963. A public relations writer organizing a campaign to promote a new model of personal watercraft, such as a Jet-Ski, would need to know about the competition. One place to look would be the Thomas Register, which contains information on who makes personal watercraft (and thousands of other products), where they are manufactured and how they are designed. A magazine writer traveling to Slovakia would be wise to search a news database such as VuText or DataTimes for the latest news stories on the country.

Internet Sources

The World Wide Web. The Internet has been described as the world's largest library, except that all the books and periodicals are piled in the middle of a room in cyberspace in no particular order. The primary way government agencies, many businesses, nonprofit groups and news organizations publish information electronically is on the part of the Internet known as the *World Wide Web*. The Web has become the vehicle of choice because of its multimedia characteristics: Sound, text and video can be integrated into one package. Web pages are extremely useful for the researcher because of *hyperlinks*, electronic pathways that send the surfer to related sites at the click of a mouse. Web indexes can help sort the clutter.

To conduct research on the Web, you need access to the Internet and a graphic browser such as Netscape Navigator or Microsoft Explorer. You also need to know how to find specific Web sites related to your topic. A *search engine* is a computer-generated index that returns a list of sites or "hits" related to the keywords you select. However, some search engines return a wider range of hits than others, so it is a good idea to use several indexes when conducting a search. Here are some popular search tools:

Subject Catalogs
Yahoo: www.yahoo.com
Infoseek: www.infoseek.com
LookSmart: www.looksmart.com
Northern Light: www.nlsearch.com

Searchable Indexes
Alta Vista: www.altavista.digital.com
Lycos: www.lycos.com
Excite: www.excite.com
Hotbot: www.hotbot.com
WebCrawler: www.webcrawler.com

Newsgroups
Deja News: www.dejanews.com

The best way to learn how to use the World Wide Web for research is by exploring. Pick a subject and start a search.

Newsgroups, Bulletin Boards and Listservs. An Internet *newsgroup* is an electronic discussion forum wherein subscribers post messages about a particular topic. Thousands of newsgroups on the Internet actively discuss topics ranging from the music of Barry Manilow to the musings of Mark Twain. The best way to find a newsgroup related to your topic is through a search engine such as Deja News, Infoseek or Excite. After subscribing to a newsgroup, you can read messages posted by others and respond to them, and post your own messages aimed at experts or people with experience in the given field. Some newsgroups are moderated; others are unmoderated. The moderated groups are preferred because the moderator helps to keep the discussion focused on the topic, whereas an unmoderated group may tend to veer off-course.

Bulletin board services, which are similar to newsgroups in that public messages are posted for perusal by anyone who signs on, also frequently provide libraries with articles, documents, pictures, software and data that can be downloaded. In addition, some bulletin boards feature a location for real-time conferences about a given subject.

Listservs are another way to observe or engage in electronic conversations about a variety of topics, but unlike newsgroups and bulletin boards, the discussion on a listserv takes place via e-mail messages that are sent to a central address. Thus, when a listserv subscriber sends a message to that address, all members of the list automatically receive it as an e-mail message. Listservs tend to be more stable than newsgroups as well as somewhat more serious. They also may be open to anyone wishing to subscribe, moderated or unmoderated, or closed to those without the desired professional credentials.

Any information acquired from the Internet, whether from a Web site, newsgroup, bulletin board or listserv, must be verified for accuracy and reliability.

Government Documents

Government documents are another useful source of information, and accessing them has become easier since the advent of the World Wide Web. There are nearly as many government document databases as there are government agencies, covering everything from patents to pollutants. Finding the right agency takes time and talent, but you should keep this guideline in mind: The information you seek exists somewhere. Keep digging until you find it.

Freelance writer and activist Bonnie Hayskar, for example, dug through a number of government Web sites to find documents on the dangers of glue-sniffing. Among the sites she visited online were the U.S. Department of Health and Human Services, the Environmental Protection Agency, the National Institutes of Health, the Centers for Disease Control and the Food and Drug Administration. She used the documents she found to show the toxic properties of toluene, an active ingredient in glue manufactured by H.B. Fuller Co. and sold in Latin America, where street children sniff it as a narcotic.[7]

Interviews

In addition to gathering information from commercial databases and Internet sources, media writers depend on interviewing as a primary way of finding information on a topic. They must be able to ask the right questions, at the right time, in the right way — with a touch of self-assurance tempered by a hint of self-consciousness. The art of interviewing is not easily mastered. It is a skill that is developed through diligence and practice. Some media writers approach the interview like Doberman pinschers, snapping out questions and demanding answers from subjects. Others are more like cocker spaniels, nudging, coaxing and cajoling.

Whatever the style or manner, interviewing depends on having the guts to ask questions — even seemingly dumb ones. For instance, while Carole Rich was a cub reporter for the former Philadelphia Bulletin, she was assigned to cover a meeting of the Philadelphia School Board. She knew nothing about the board or its business, but she noticed this little item on the agenda: "Approves token losses of $30,000." After the meeting, she got up the courage to ask a school auditor the obvious question: How can $30,000 be a token loss? He told her it was $30,000 worth of lost bus tokens that the school district sold to students to use on the city's public transit system. Rich, in turn, asked more questions of other officials, eventually learning that a major theft of tokens had occurred at one of the schools and that the procedure for selling tokens varied from one school to another. The auditor, who did not want to be quoted, supplied Rich with facts and figures about previous years' losses. Rich explains what happened next:

> I returned to the newspaper and told my editor about the tokens, the only item that really interested me in what seemed like an otherwise boring meeting.
> "Write it," he said. And he told me to put the rest of the board's news in a separate story.
> The next morning, I awoke to find my token story stripped across the front page. And the next day the school board announced that it would devise a uniform policy for selling the tokens at all its schools. After that, I was assigned to cover the education beat, one of the best beats on any newspaper.[8]

What Carole Rich got by asking the right questions, at the right time, in the right way, was a good story and a promotion. Not all media writers can expect such dramatic results, but they can expect the answers to good questions to make good stories easier to write.

Locating Interview Sources. The key to getting good answers to good questions is finding the right person to ask: one who has the knowledge, time and inclination to talk about your topic. Sometimes, the media writer needs only facts, perhaps about an influenza virus attacking grade-schoolers; here, the task is finding the appropriate agency, such as a city or state public health department, and a specialist in the communicable diseases division. Other times, the media writer searches for the answer to a policy question, such as why clinics did not have adequate flu vaccine supplies for everyone in the community; here, the writer needs to find someone higher up the administrative ladder who can speak for the agency or department. The secret is in finding a bureaucrat who is important enough to be unafraid to talk but not so important as to be unavailable.

Putting together the best story, however, requires more than quotes from public officials. A story about a flu outbreak would be a pale imitation of real life if it did not have a few quotes from a child with an aching head, upset stomach and sweaty brow. The child's mother also might have a strong quote or two about what a virus can do to a household routine. Good interviewers, then, get the official statement about a situation and go on to explore its colorful consequences for ordinary people. For instance, after talking with the academic dean about a college's new grading policy, the interviewer will want to find the students who are suffering its effects.

The best interviewers are honest and straightforward. They state the purpose of the interview, identify the information they seek and describe its role in the story they plan to write. They also respect their sources' time, avoiding unnecessarily long interview sessions and keeping appointments that have been set aside for interviewing.

Identifying Types of Interviews. Interviewing — the art of seeking information from people through questioning — is a lot like living life: Sometimes it is hard, sometimes it is easy and sometimes the fear is worse than the reality. How a particular interview goes depends on the skill of the interviewer and the attitude of the subject. It also depends on the kind of interview and the circumstances surrounding it.

Media writers can expect to encounter several types of interviews in their work. These include the informational, situational, confrontational, personal and professional types of interviews.

The Informational Interview. The most simple and straightforward type of interview situation is the informational interview. Here, the interviewer seeks information (such as facts and figures) about a program, policy, product, procedure or point of view from an expert who is willing to share his or her knowledge with the interviewer. Most print and broadcast reporters spend a part of each working day seeking to be informed and educated by experts who have more knowledge than they do about the subjects of their stories. Similarly, public relations writers turn to company executives and managers for the details of a new policy or program they are writing about in a news release or an in-house newsletter. And advertising copywriters attend meetings

WRITING TIPS

USING QUOTATIONS EFFECTIVELY

Interviews are a rich source of quotations. In a print or television news story, an advertisement or a press release, quotations can add zest and color to the writer's words and give readers or viewers a break from the usual pace and structure of the story. A good quotation functions like a verbal snapshot, explaining or expanding upon some element of the story.

The following guidelines will help you use quotations effectively in your writing:

1. Use quotations in complete sentences. Quotations that appear as full sentences are easier to read and understand. Taking good notes during interviews will help you develop an ear for good quotations and the skill of recording them in full-sentence format.
2. Choose quotes carefully. Cluttering a story with mediocre quotes will diminish the power of the effective quotations in a story.
3. Highlight an especially strong quotation by presenting it in a separate paragraph.
4. Avoid "stacking" quotes. These are quotations from two different speakers stacked in back-to-back paragraphs. Introduce the second speaker with a new paragraph by paraphrasing, for example, something he or she has to say.
5. Give the attribution at the end of a quotation for stories that will appear in print. For example:

> "The information is often more important than the name of the person who relayed it," she said.

When the quotation is two or more sentences in length, place the attribution after the first sentence:

> "I like ice cream better than frozen yogurt," John Smith said. "But I'd rather have frozen yogurt than no dessert at all."

6. Word the attribution according to the context in which it is used. Usually, *said* comes after the person's name or the referring pronoun, as in the two preceding examples. However, note that *said* comes before the name or pronoun when you describe the person being quoted:

> "I like ice cream better than frozen yogurt," said Jim Smith, who owns a chain of ice cream stores. "But I'd rather have frozen yogurt than no dessert at all."

7. Use *said* or *says* consistently in the attributions given throughout a story, avoiding inappropriate shifts in tense. The present-tense *says* is sometimes used in feature stories. Also avoid substituting the neutral attribution word *said* or *says* with a non-neutral word such as *exclaimed, announced* or *sighed.*

with agency marketing and research experts to get information about consumer tastes, product qualities and brand loyalties.

When television reporter David Wildermuth seeks information through interviews, he says he casts the net as far and as wide as possible, even including people he thinks might not have relevant information. In 1995, for example, Wildermuth was sweating a deadline as he put together a story about a TV photographer who claimed he was roughed up by the chief of Minnesota Vikings security and an off-duty police officer during the halftime of a "Monday Night Football" game. "When you get a story like this and don't have much time, you have to figure out who the players are and what you need to make the story flow," Wildermuth says. "And then you have to make phone calls and ask questions." The reporter called the Minnesota Vikings first, and as he recalls, got a predictable nonresponse. Next, he interviewed the police officer to get his version of the scuffle. Wildermuth learned that the photographer was being verbally abusive and "acting like a jerk," according to the officer. The interview was short, and it was not conducted on camera. But the few words from the police officer added some needed perspective to the story. "You can never assume someone won't talk," Wildermuth says. "You might be surprised. I've been turned around on stories. I suspect things are one way, and then somebody puts a nugget in my head. I say, 'I can see that point, too,' and the story takes another turn."[9]

The Situational Interview. Situational interviews are typically initiated and governed by the immediate circumstances, requiring the interviewer to think and act quickly, with subtlety and sensitivity. Media
writers are most likely to conduct a situational interview at the scene
of an accident, a murder or a fire, or in the aftermath of a natural disaster such as a hurricane or a flood.

Many beginning media writers may dread having to interview a victim's family or the survivors of a tragedy. This type of interview can be
intrusive and invasive, delving into a person's private life. In some cases,
it is best for the reporter to avoid asking questions and for the photographer to put down the camera. At other times, however, a thoughtful
questioner can elicit answers without asking sensitive questions directly.

Consider, for instance, how one rookie reporter working the night
shift for a Sacramento television station handled a situational interview with the family of a young woman who had been murdered.
Upon arriving at the family's home, Joann Lee identified herself, apol-
ogized for intruding and asked to speak with a family member. She
was escorted by the victim's uncle into the living room where about
two dozen people — aunts, uncles, siblings, friends and parents —
were gathered in silence. But soon they began to talk, and as they did,
the pieces of Lee's story began to take shape. Gradually, the victim became much more than a statistic in a police report. She was an 18-
year-old woman with dreams and aspirations and a loving family. Her
family, too stunned, angry and hurt to talk about the crime, wanted to
talk about her in positive terms so that when the young woman's name
was broadcast on the evening news, she would be portrayed not only
as a victim of a brutal crime but also as a person with a life, a family, a
future.[10] The photographer shot the victim's graduation picture, the
family sitting in the living room and the uncle talking about the victim's hopes and plans. And Lee got her story without having to ask
those dread questions.

Public relations writers also have to perform situational interviews.
In a crisis situation, for example, the PR person for a company interviews the people involved in the accident, fire or other crisis. The information is then passed on to inquiring reporters, company employees or people who live near the company.

The Confrontational Interview. A confrontational interview has the
hardest edge and the biggest stakes for the interviewer and interviewee
alike. Its purpose is to ask specific questions and to get specific answers, but media writers can find themselves on either side of the

questions, asking or answering them. A public relations vice president may be responding to a reporter's question about a sexual harassment complaint filed by several employees. Or a police reporter might be asking the chief why less than 5 percent of reported allegations of police brutality are substantiated by department investigators.

For Teresa McFarland, director of public relations at Mall of America, the nation's largest indoor shopping and entertainment complex, asking questions is a crucial part of her job, particularly when she expects to be confronted by reporters. "As a public relations person, I do a lot of seeking answers — to questions that I expect reporters to ask," McFarland says. While preparing a news conference to announce Mall of America's parental escort policy for children under age 16, for example, McFarland thought about all the questions reporters might ask, and then she turned to mall executives and merchants for the answers. As she explains, "The press conference went very well. I don't think the reporters asked a single question that we hadn't anticipated. A good PR person always has to put on a reporter's hat. That's what I did — put on my reporter's hat. I try to be skeptical and ask the toughest questions I can think of."[11]

The people involved in a confrontational interview need to be especially well prepared as well as thoughtful in how they ask and answer questions. They also may need to be assertive. Consider Barbara Walters, who has been described as "the best-known woman journalist on TV chiefly because of her tough approach to interviewing."[12]

> After Walters asked Mamie Eisenhower if she was aware of the longstanding rumor that she was a dipsomaniac, friends asked the interviewer "How could you have asked?" Says Walters: "I find very often people like to confront rumors. It depends on how much they trust you. And you have to have a line between what is tasteful and what isn't." (In the case of Mamie Eisenhower, it turned out the former First Lady had an inner-ear infection that made her appear woozy on occasion.)[13]

The Personal Interview. The personal interview is the foundation of the personality profile, wherein the interviewee is the focus of the story. The writer wants to evoke feelings and emotions with the interview questions and thereby gain insight into the character's personality. The profile could be a newspaper feature about a local baseball team's public address announcer who happens to be blind, or it could be a cover story about a new CEO for a company magazine. Whomever it is about, the profile should brim with colorful anecdotes, revealing observations and lively quotations gleaned from the interview. But before media writers can conduct a personal interview they need to do some research into the subject's life and work.

In 1993, for instance, reporter Renee K. Gadoua was assigned to write a profile of Ellen Tarry, an 88-year-old African American writer and activist. Before Gadoua prepared the questions she would ask Tarry, she read Tarry's autobiography, tracked down articles she wrote and found references to the writer in several books on the Harlem Renaissance. As the reporter explains, the research she conducted gave her a rich picture of her subject's life and accomplishments:

> As my notes piled up, a fascinating story began to emerge. Ellen Tarry was born in Birmingham, Alabama, and was raised a Congregationalist. Her parents, both of mixed blood, sent her to a Catholic boarding school in Virginia, sparking an interest in Roman Catholicism that led to her conversion and baptism in 1922 at the age of seventeen.
>
> During Tarry's career as a journalist, her writings frequently decried injustice and segregation. She later moved to New York City, where she found herself in the midst of the Harlem Renaissance, the revival of black culture and literature. Among her friends were the celebrated poets Claude McKay and Langston Hughes.[14]

By the time Gadoua hopped on a train to New York to interview Tarry, she knew enough about her subject to ask the kinds of questions that would reveal Tarry's character and conscience. Her preparation paid off. Gadoua describes the payoffs in this way: "I had expected [Ellen Tarry] to spare an hour — two at the most — to talk to me. Instead, the interview lasted all afternoon and included poring over scrapbooks and family pictures."[15]

The Professional Interview. The professional interview differs from other types of interviews in that it takes place in front of an audience, rather than in a one-on-one private setting. Examples include interviews conducted on a television newscast or talk show, at an annual stockholders' meeting or with a focus group for marketing executives. The interviewer — whether on camera, at the podium or at the head of the table — attempts to do several tasks simultaneously: to ask precise questions, elicit informative answers, interrupt or redirect subjects whose responses are long-winded or off-topic, control the focus of the conversation, keep an eye on the clock and come up with a good getaway line when it is time to end the interview. The idea is not to bore audience members but to ask questions they want answered.

Professional interviews often rely on open-ended questions. Such questions, designed to prevent subjects from giving simple yes/no responses, aim to keep the conversation flowing. The interviewer wants subjects to talk at length about the topic at hand, whether it is an accident they witnessed, a controversial bill they proposed or a breakfast cereal they sampled.

Observe General Guidelines for Interviewing. No set of techniques will guarantee a successful interview. However, by observing the following guidelines you can increase your chances of conducting fruitful, productive interviews, whether you are a veteran reporter talking to a savvy city council president or a public relations writer drafting a speech for a company vice president.

Treat the interview seriously. Remember, the interview should focus not on casual conversation but on extracting the details you need to write the story.

Prepare for the interview. At times, you may have only a few minutes to gather your thoughts before talking with the owner whose store is burning, for example. Use every minute you do have to prepare for the interview. Lack of preparation can terminate the interview, leaving you with nothing to write about. Consider this example:

> When Vivien Leigh arrived in Atlanta for the premiere of the reissue of *"Gone with the Wind,"* a reporter asked her what part she had played in the film. Scarlett informed the writer that she did not care to be interviewed by such an ignoramus.[16]

Establish a relationship with the subject. This does not mean becoming best friends. But it may mean asking the subject about the family pictures on the desk. The point is simple: People will talk more candidly if they like you, or if they think they might like you if given the time.

Sports columnist Mitch Albom of the Detroit Free Press tells us how he established a friendly relationship with his subject, Mel Bridgman, a hockey player from Philadelphia who came to Detroit.

"First thing I said to him was, 'You're from Philly?' And he said, 'Yeah.' I said, 'Soft pretzels.' And he said, 'Yeahhhh.' Then we traded thoughts about where the best pretzels were sold on the street in Philadelphia. By the time I got around to start interviewing the guy, we were chums. You take a little fact like that into an interview and you can open a hundred doors."[17]

Clarify the ground rules for the interview. If the entire interview is to be on the record, make that clear to the source *before* asking the first question. The understanding between the interviewer and the source may, and has, become the focus of lawsuits.

Ask brief and specific questions, not long-winded or vague ones. You want to get answers, not hear yourself talk. Avoid asking multipart questions: They can easily confuse your subject and prevent you from getting the information you need.

Give the interview subject enough time to reply. Allow time for silence, if necessary. For some subjects, silence may precede a thoughtful, emotional answer.

Listen to those answers. Although you have prepared a set of interview questions, you need not stick doggedly to the game plan. When your subject replies to a query with a surprising declaration, avoid plowing ahead to the next question instead of reacting to the subject's last answer. Try to be as flexible as possible, as Tom Wheeler was in his interview with B.B. King (see pp. 77–78). Startling information can often emerge from an interview that is allowed to take its own course.

Follow up on vague answers. Ask your subject for more evidence, more details, more candor. Read back the answers to be sure your subjects understand what you think they have said.

Encourage your subjects to express themselves freely, especially those who are not experienced in dealing with the media. You should neither suggest a response nor offer an opinion.

Use a mix of open-ended and closed-ended questions. Each type of question has its place in an interview. The closed-ended question asks of the council president: "Will there be a property tax increase?" The open-ended question asks: "Why?"

Observation

In addition to being successful information gatherers and interviewers, good media writers are keen observers. At a campaign rally, a news reporter looks for candidates' actions he or she can describe to

WRITING TIPS

LEARNING ABOUT TYPES OF INTERVIEW SUBJECTS

Seasoned reporters know that no two interview subjects will be alike. Some will sit at mahogany desks wearing monogrammed shirts and designer loafers; others will be people who work with their hands, clad in blue jeans and steel-toed boots. When he was teaching at the University of Wisconsin, Professor Steven Chaffee told students that reporters should recognize at least three types of interview subjects and adapt their interview questions and general approach accordingly:[1]

1. *Potentates* are people in positions of power. She can be a governor, he can be a police chief, or they can be on the board of directors. Whoever they are, they can control the interview, intimidate the interviewer and even evade the tough questions. In this situation, the interviewer had better come armed with a list of smart questions, asked clearly.
2. *Experts* are among those most frequently interviewed by media writers; they may include scientists, plant managers, union presidents and computer technicians. They know more about their particular areas than most readers or viewers need to know. The interviewer's task is to make sure he or she understands the answers to questions and can translate them into clear and simple prose.
3. *Just plain people* sometimes become targets of interviewers' questions. They may be victims, witnesses, bystanders, fans, shoppers or passersby. And they ought to be treated respectfully by the interviewer, who clearly states the purpose of the interview, explains the ground rules and encourages candid responses.

1. Steven Chaffee, "The Interview as a Reporting Tool," handout for Journalism 203/204, 1966.

his or her audience. At the building site, a public relations writer observes carpenters to write a media kit about a housing development. At the store, an advertising writer watches shoppers' behavior to glean information for writing a series of ads.

For example, when the account planning team from the advertising agency Campbell Mithun Esty began working on its strategy for Domino's Pizza ads, team members temporarily became pizza deliverers. "Our account planners literally rode with the Domino delivery people to homes and watched the reaction when the pizza arrived," said Howard Liszt, chief executive officer of Campbell Mithun Esty. "We learned points of view that traditional research would never have uncovered. Children reacted as if Santa Claus was coming through the front door. It wasn't just pizza. That information became the bedrock of our strategy."[18]

Television reporter Dennis Stauffer also is a keen observer, which helped him tell a livelier story the night Mall of America implemented its escort policy in 1996, requiring children under age 16 to be accompanied by an adult on weekend nights. "The night this thing goes into effect, I'm to look to see how things go," Stauffer said. "I also want to spot the bugs, the things that don't work so well." With those goals in mind, Stauffer began wandering from one mall entrance to another, when he spied a 20-year-old woman in a wheelchair hassling with security guards who claimed her 15-year-old aide could not accompany her into the mall. The hassle ended when a bystander offered to escort the wheelchair-bound woman and her aide. Stauffer and his photographer captured the exchange on videotape, including the moment when a mall manager agreed to make an exception for the woman and her aide. "This is one of those little nuggets you're hoping to find when you put together a story," Stauffer said. "I didn't need to even ask a question. I just spotted the moment and let it happen."[19] The happening added spice and spirit in the middle of his videotaped story, which you will read more about in Chapter 6 (pp. 142 and 144).

SORTING AND ORGANIZING INFORMATIONAL MATERIALS

Once writers gather their material — facts and statistics from documents, ideas and quotations from interviews, anecdotes and details from observation — they must organize it. A basic story tends to be organized naturally by providing information that expands on the most important facts given in the opening. But careful organizing is crucial in all media writing, especially in more complex writing, where the structure is not necessarily determined by the facts.

Most writers begin organizing as they gather material. For instance, they may place a checkmark or star by a particularly strong quotation or example as they record it in their notes. A writer working on a multi-interview story may notice an emerging theme and focus subsequent interviews and research more narrowly on that theme. Even when writers identify possible themes during the interview process, all the gathered material still must be organized and sorted before writing begins. Some writers read through their notes, marking the margins to identify important quotes, facts and perspectives. Then they go through the notes again, using color-coded markers to highlight related material by theme or subtopic. Other writers, particularly print and broadcast investigative reporters and those who write 3,000- to 7,000-word magazine articles, develop more elaborate systems of organization. They may create binders or folders in which notes, observations, documents and photocopied articles are classified and filed, or they may create computer files where they store notes and research.

Bruce Bendinger, an award-winning advertising copywriter and creative director, says that after he accumulates "piles of notes, random scribblings, memos, magazine articles, competitive ads and product literature," he then organizes the material in a logical manner, generally by outlining.[20] That's what most writers of complex or in-depth material do once they have made sense of what they collected: They prepare an outline or a brief sketch of the article's probable major sections.

THE NEXT STEP: WRITING THE OPENING

Bendinger notes that gathering information is a difficult but critical task for media writers:

> Before you can write clearly about a subject, you must understand it.
> It means assembling information.

Digging for facts — the preparation stage of the creative process.

Much of the difficulty of not knowing what to say is rooted in not having anything to say.[21]

Gathering information from online and print documents, databases, interviews and observation gives media writers something to say. They then begin writing by using those details and information in a way that will capture the audience's attention. The first few sentences of any news story, advertisement or news release are the key to capturing readers' attention. These first sentences, the *opening* or the *beginning* of a piece of writing, are our focus in Chapter 5.

NOTES

1. Tom Wheeler, "When in Doubt, Ask!" in *Journalism: Stories from the Real World*, ed. Retta Blaney (Golden, Colo.: North American Press, 1995), 30–31.
2. Ibid., 31.
3. Shirley Biagi, *Interviews That Work: A Practical Guide for Journalists* (Belmont, Calif.: Wadsworth, 1986), 31–32.
4. Quoted in Denis Higgins, *The Art of Writing Advertising: Conversations with William Bernbach, Leon Burnett, David Ogilvy and Rosser Reeves* (Chicago: Advertising Publications, 1965), 38.
5. Jean Ward and Kathleen A. Hansen, *Search Strategies in Mass Communication*, 3rd ed. (New York: Longman, 1997), 92.
6. Nora Paul, *Computer Assisted Research: A Guide to Tapping Online Information*, 3rd ed. (St. Petersburg, Fla.: Poynter Institute, 1996), 22.
7. Bonnie Hayskar, personal communication via e-mail, 11 July 1997.
8. Carole Rich, "When a Dumb Question Leads to a Great Story," in *Journalism: Stories from the Real World*, ed. Retta Blaney (Golden, Colo.: North American Press, 1995), 29.
9. David Wildermuth, personal telephone interview from Minneapolis, Minn., 25 June 1997; "Scuffle," KARE 10 p.m. news, 31 Oct. 1995.
10. Joann Lee, "How Do You Feel About That?" in *Journalism: Stories from the Real World*, ed. Retta Blaney (Golden, Colo.: North American Press, 1995), 146.
11. Teresa McFarland, personal telephone interview from Bloomington, Minn., 26 June 1997.
12. John Brady, *The Craft of Interviewing* (New York: Vintage Books, 1976), 90.
13. Ibid., 90.

14. Renee K. Gadoua, "Knowing What *Not* to Include," in *Journalism: Stories from the Real World*, ed. Retta Blaney (Golden, Colo.: North American Press, 1995), 38.
15. Ibid., 39.
16. Brady, *Interviewing*, 36.
17. Quoted in John Sweeney, "The Profile, the Interview," *Coaches' Corner*, Dec. 1993, 4.
18. "Executive Focus: Howard Liszt, Agent of Creativity," *St. Paul Pioneer Press*, 12 Oct. 1997, 2D.
19. Dennis Stauffer, personal telephone interview from Golden Valley, Minn., 25 June 1997.
20. Bruce Bendinger, *The Copy Workshop Workbook* (Chicago: Copy Workshop, 1988), 151.
21. Ibid., 150.

5

WRITING THE OPENING

*G*ot Milk?

— California Milk Processor Board ad[1]

All types of media writing should have an effective opening, one that captures and holds the interest of the audience. In 1993, when the California Milk Processor Board wanted to reverse a decade-long decline in milk consumption in the United States, it turned to the San Francisco advertising agency Goodby Silverstein & Partners. The agency used humor to capture the attention of readers and viewers and to pound home the idea that milk is a necessity, especially when eating baked goods like cookies and brownies and palate-sticking peanut butter sandwiches. "Got Milk?" was used as a headline in print ads and as a slogan in television ads. As an attention-getting statement, it was concise, clever and memorable. And it seemed to work.

Coupled with the "Milk: Where's *Your* Mustache?" campaign (see Fig. 5-1), which was initiated in 1995 by the National Fluid Milk Processor Promotion Board, the "Got Milk?" campaign helped volume sales of fluid milk rise in the first three quarters of 1996, according to figures from the U.S. Department of Agriculture. In California, the consumption of milk stabilized at 23 gallons per person annually. "We consider stabilizing it to be reasonably successful for us," said a spokesperson for the California Milk Processor Board.[2]

CAPTURING THE AUDIENCE'S ATTENTION

The crafting of the first words in a story is one of the most important tasks of the media writer. Each story must have a beginning, a

http://www.whymilk.com

What's a hip drink?
Let me lay some lyrics on you,
man. Hey, *Chocolate* milk.
It's nutty, it's too neat. Get your
nutrients with a koo-koo beat.
Get it lowfat, get it fat free.
It's a swingin' chocolata treat.
Chocolate milk, oh yeah!

MILK
Where's *your* mustache?"

Figure 5-1

middle and an end, but readers choose whether to spend time with a story not by the content of the middle or end but by the effectiveness of the beginning. An opening can be as simple as "Got Milk?" or as complex as a trade magazine description of a product, as long as it tells readers what the story is about and promises more to come. Thus, a good beginning should serve a dual purpose: to interest the reader and hold the reader's attention. These goals are not as easy to accomplish as they may sound.

Rene J. Cappon, an Associated Press editor and author of "The Word," stresses the difficulty and significance of writing good introductions: "Think of them as though they cost you 10 bucks per word, each word to be engraved on stainless steel while you're sitting on a hot stove. Think economy."[3] Cappon advises writers to look for what is different in a story because that is where the beginning might be hiding. Sometimes the best way to come up with a starting point is to tell the story verbally to a friend. Often the first words out of your mouth will be the most important and the most interesting aspects of the topic.

USING ATTENTION-GETTING STATEMENTS AS LEADS

Any type of media writing has a beginning, middle and end, whether it is an advertisement for milk, an industry magazine article on dairy cattle management or a television news story on brucellosis. Most writers for the mass media refer to the beginning as the *lead* or *lead-in* (formerly spelled *lede* to distinguish it from *lead type*). Among the most memorable leads in radio broadcast history is CBS's Edward R. Murrow's classic World War II opening: "This is London," as bombs whistled past during the Battle of Britain in late 1940. Murrow's reports, which were credited with stirring the American people to aid the Allies, prompted the poet Archibald MacLeish to recall: "You burned the city of London in our houses and we felt the flames that burned it."[4] When war came to Americans, slightly more than a year later, President Franklin D. Roosevelt, in a speech before Congress, began with another great opening line: "Yesterday, December 7, 1941, a date which will live in infamy. . . ."[5]

Although public relations professionals and advertising copywriters deal with crises of a different sort, they too strive to write effective openings. For the public relations writer that often means using straightforward language. For example, during the apple pesticide Alar scare of 1989, the interest group Mothers and Others for a Livable Planet began its fact sheet with a clear message: "Our children are safer with Alar off the market."[6] Similarly, when Target Stores decided in 1996 to stop selling cigarettes, its news release began with this straightforward statement: "Target is discontinuing the sale of cigarettes in our stores."[7]

Advertising copywriters deal with visual images as well as text, but they are advised to remember ad guru David Ogilvy's famous statement, "Advertising is a business of *words*." Look again at Figure 5-1. The "Milk: Where's *Your* Mustache?" campaign featured a number of celebrities, such as Lauren Bacall, Kate Moss, Pete Sampras and Spike Lee, wearing a mock milk mustache on their upper lips. Created by Bozell Worldwide in New York for the National Fluid Milk Processor Promotion Board, the agency's first ads carried the headline "Milk: What a Surprise!" That phrase failed to reinforce the visual pun of the milk mustache, and it eventually was replaced in the second phase of the campaign. Along with the slogan change, Bozell Worldwide added a reference to the Web site (http://whymilk.com) and expanded what was a magazine-only campaign to include other media formats such as billboards and transit posters.

Some of our culture's most enduring phrases originally appeared as openings in advertisements. For instance, "Often a bridesmaid but never a bride" was written in the 1920s as the lead for Listerine mouthwash ads. More recent examples include the jingle "This Bud's for You" (Budweiser beer) and the well-known humorous question "Where's the Beef?" (Wendy's), both of which have become part of the national lexicon.

USING LEAD-INS WITH IMAGES

Avoiding repetition is especially important when crafting a lead-in that will be used with images. In broadcast news, the lead-in is the anchor's introduction to a story, which is followed by the reporter's videotape. In the following example from a 1996 KARE newscast, the lead-in to Brad Woodard's plane crash story begins with the anchor, turns to the reporter, and then moves to the reporter's videotape:

ANCHOR: Federal investigators are in north central Minnesota tonight trying to uncover the cause of a plane crash that killed four people.

The victims took off from International Falls yesterday afternoon in a high-performance Piper Malibu.

They soon flew into thunderstorms and hail, just before 2:30. The plane went down near the town of Aitkin. KARE-11's Brad Woodard joins us with the latest.

REPORTER: Pat, the men in the plane were successful businessmen from Ottumwa, Iowa, returning from a fishing trip.

The Piper Malibu was flying at 17,000 feet when pilot Joseph Carpenter Jr. radioed that he was in trouble, that he may have lost his tail.

Local farmers fought their way through brush, water and trees on A-T-Vs. The sight of the twisted wreckage will stay with them. *(Then the videotape story begins.)*[8]

The story's video images start with a sound bite from a farmer who says, "The plane was just ripped apart." The lead-in is clear and tight: The first paragraph says four are dead in a plane crash, and each succeeding paragraph adds more detail until the last paragraph, which makes a smooth transition to the opening sound bite of the videotape story.

IDENTIFYING COMMON TYPES OF LEADS

Ring Lardner, the famous writer, penned what many journalists consider to be the best newspaper lead ever written. About the death of a promising young boxer, Lardner's lead included this sentence: "Stanley was 24 years old when he was fatally shot in the back by the common law husband of the lady who was cooking his breakfast."[9] The information conveyed in that lead goes beyond 26 words on the printed page.

Lardner's opening is an example of a summary lead, the most common type of lead used by media writers. Several other types of leads are also used in media writing, including the multiple-element, suspense, character, scene-setting and narrative leads.

The Summary Lead

The main function of a *summary lead* is to provide information. As the term implies, this type of lead summarizes the important details of

WRITING TIPS

Observing Traditional News Values

Traditional news emphasizes six values.[1] Journalists consider these values in choosing events to cover, and public relations writers stress the values in pitching stories with their clients' messages to journalists.

1. *Impact:* An event's consequence for the audience affects its news value. A 16-car pileup stalling 600 expressway commuters is bigger news than a fender-bender on a rural road.
2. *Proximity:* Where the event occurs is important, and events closer to home are bigger news. Therefore, a 16-car pileup in Denver is important for Colorado TV stations but not for Atlanta stations.
3. *Timeliness:* Newer news is bigger news. If that 16-car pileup occurs today, it is a bigger story on tonight's newscast than a car crash of three days ago.
4. *Prominence:* Simply put, names make news. That means a car crash that kills the governor is bigger news than a crash that kills a garbage collector or a college professor.
5. *Novelty or deviance:* Unusual things make news. A car crash caused when a driver swerves to avoid an elephant that escaped from the zoo is bigger news than a crash caused when a driver falls asleep.
6. *Conflict:* Contention between people or organizations makes news. If the police investigating the 16-car pileup draw different conclusions about its cause than the driver of the first car, the story will continue to make news.

1. The Missouri Group: Brian S. Brooks, George Kennedy, Daryl R. Moen and Don Ranly, *News Reporting and Writing*, 5th ed. (New York: St. Martin's Press, 1996), 3–4.

a news event. Typically, summary leads are one sentence long, although there is no rule against two short sentences or more. Summary leads are used in "hard" news stories written on deadline, shorter stories, routine news and news releases. Sports stories, business news, weather stories and news releases often use summary leads. Media consumers in a hurry or those who skim or channel hop can get the gist of the event from a summary lead. Entire ad campaigns can be based on what is essentially a summary lead: "Drink Coca-Cola" or "Timex: Takes a Licking and Keeps on Ticking." As news and other information becomes more compressed for various audiences, summary leads will only become more ubiquitous.

Six elements are common to summary leads: *who, what, where, when, why* and *how*. They can be stated as questions:

What happened?
Who was involved?
Where did it happen?
When did it happen?
Why did it happen?
How did it happen?

Most summary leads contain at least four of the five *W* questions and the *H*. In such a lead, the *W*s and *H* are usually mentioned in order of importance, which depends on the subject matter and the medium. For example, a newspaper story would rarely begin with a *when*, or time element; a broadcast story, on the other hand, might start out, "This morning at the White House, President Clinton. . . ." The immediacy needs of a broadcast audience help dictate beginning a lead with *when*.

For most news stories, the *what* or *who* elements are most important. The *what* element represents the news value in the event — the action. That is why the *what* element often is stated as a verb. For stories involving famous people, the *who* element becomes the most significant. Almost any action taken by the U.S. president, for example, is considered newsworthy. *Where* and *when* — which are not space-consuming in the crowded world of lead construction — are usually subservient to *what* and *who*. The *why* and *how* elements are sometimes left out.

Here is an example from a team of reporters at the Chicago Tribune, covering an airplane accident under a tight deadline in November 1996:

QUINCY, Ill. — A United Express plane from Chicago and a private plane collided on a runway in a fiery accident Tuesday night that killed 13 people.[10]

The elements of the lead can be dissected as follows:

WHO: A United Express plane and a private plane
WHAT: collided, killing 13
WHERE: on a runway in Quincy, Ill.
WHEN: Tuesday night

The *how* element was left for the second and third paragraphs in the story, and why the planes crashed was not known at press time and was not included.

Obituaries often open with summary leads. When the retired commissioner of the National Football League died in December 1996, reporter Bill Brubaker of The Washington Post used a standard, albeit fact-filled, summary lead:

> Alvin "Pete" Rozelle, a masterful promoter, innovator and deal-maker who was commissioner of the National Football League for almost 30 years until his retirement in 1989, died last night at his home in Rancho Santa Fe, Calif. He was 70.[11]

A reader skimming the newspaper on that December morning could read the first paragraph of Brubaker's obituary and get the principal facts:

WHO: Pete Rozelle
WHAT: died
WHEN: last night
WHERE: at his home

Rozelle was a familiar name to many, but not all, Americans. To jog the memory of some readers and introduce the subject to the rest, the writer uses a description of Rozelle's style and job. The cause of death — *how* — in Rozelle's case, brain cancer, is mentioned in the second paragraph. The *why* of someone's death is a question left for philosophers, theologians and scientists and generally is not included in newspaper obituaries.

Summary leads are not limited to newspaper stories. Public relations writers are well advised to present facts and figures in clear, concise, no-nonsense terms in a news release. That way, their audience — journalists — can withdraw the news from a news release in a hurry. Here is how Teresa McFarland, public relations director for Mall of America, announced the Mall's new parental escort policy:

> BLOOMINGTON, Minn. — Mall of America will implement a Parental Escort Policy effective Oct. 4, 1996, to reduce the growing number of unsupervised youth at the Mall on weekend nights.
>
> Under the new policy, youth under 16 will need to be accompanied to the Mall by a parent or guardian 21 years or older, from 6 p.m. until closing time on Friday and Saturday nights. Youth under 16 who do not have a parent or guardian with them will not be allowed to enter or remain in the Mall after 6 p.m.[12]

McFarland uses simple and direct language to state the message. A newspaper or broadcast outlet could use her first sentence with only minimal changes. One pitfall for public relations professionals — avoided by McFarland — is a tendency to bury or sugarcoat the news; such an effort damages the credibility of the PR person in the eyes of the audience.

Writers for the World Wide Web are presented with a unique set of circumstances when composing the opening for a Web page. The World Wide Web is useful to public relations writers in distributing news releases quickly, and a summary opening, complete with electronic-mail connections and other links to supporting materials, is a typical strategy. Here is a University at Buffalo release about a scientific study conducted by its faculty:

> BUFFALO, N.Y. — Eating contaminated sport fish from Lake Ontario is associated with shortened menstrual cycles, epidemiologists from the University at Buffalo have found.[13]

Then, next to the story are eight links to other resources, including experts in the field.

Let us look at a more complex example of a summary lead. In January 1997, Newt Gingrich, speaker of the U.S. House of Representatives, was involved in a controversy surrounding an ethics investigation at the same time that he sought re-election as speaker. Here is how USA Today reporter Jessica Lee, using a summary lead, begins her story on the results of the voting for speaker:

WASHINGTON — Newt Gingrich triumphed Tuesday over Democratic opposition and Republican dissent to become the first Republican in 68 years to win successive terms as House speaker.

Gingrich, his re-election clouded until the last minute by ethics charges, won 216 votes — three more than he needed.[14]

There are only 25 words in the first paragraph (which, in its pure form, could be considered the lead) and 20 words in the second. Breaking down the elements, here is how it looks:

WHO: Newt Gingrich
WHAT: wins re-election as speaker
WHERE: U.S. House, Washington
WHEN: Tuesday
WHY: by triumphing over opposition and dissent
HOW: won 216 votes

Note that Lee fits the *who, what, where, when* and *why* in the first paragraph, and adds the *how* in paragraph two, along with more explanatory material about the ethics charges. Her first words are *Newt Gingrich*, because as speaker of the house (and a well-known national figure), Gingrich is the central focus of the story. Also notice that the time element, *Tuesday*, is mentioned immediately after the strong main verb, *triumphed*. A good rule of thumb is to get the time element after the main verb unless the day of the week can be mistaken as the object of the verb. "George Foreman fought Tuesday . . ." isn't going to work, unless a fighter out there is named after a day of the week. One more note: A time element tacked on to the end of a sentence too often looks like an afterthought.

Immediate Versus Delayed Identification Summary Leads. Some summary leads are divided into two categories: immediate identification and delayed identification. *Immediate identification leads* always mention a person by name — the emphasis is on the *who* of the story. This type of summary lead is used when the person making the news is the most important aspect of the news.

In a *delayed identification lead*, the emphasis is on the *what* of the story. This kind of lead is used when what happened is more important than the person to whom it happened, such as when the person or persons involved in the news have little name recognition beyond their immediate circle. In a story on a traffic accident involving a person not in the public eye, the reporter would use a delayed identification lead, substituting a brief descriptive phrase in place of the person's name. It then becomes important to mention the name as soon as possible.

DES MOINES, Iowa — A Bloomington, Iowa, man died Sunday night when his car collided with a deer on Interstate 80 near Ankeny.

John Doe, 70, was killed instantly when the animal smashed through the windshield, the Polk County medical examiner said.

The phrase *Bloomington, Iowa, man* is used to describe the victim, who is then introduced by name in the second paragraph. If the accident involved a well-known person, the immediate identification lead might read:

DES MOINES, Iowa — Former Iowa governor John Doe died Sunday night when his car collided with a deer on Interstate 80 near Ankeny.

Doe, 70, was killed instantly when the animal smashed through the windshield, the Polk County medical examiner said.

News releases from big corporations will almost always use immediate identification leads, assuming the company is well known. For example, a "pitch letter" to newspaper and magazine editors from a PR agency hoping to get a story on its client begins this way:

> In an industry where many companies are folding or cutting back, IDS Financial Services is growing and prospering.[15]

There are no set rules on when to use summary leads or any other type of lead. And summary leads will differ from medium to medium. In broadcast writing, where one rule of thumb in lead writing is to

WRITING TIPS

USING ACTIVE VERBS

Use active verbs. That is probably the single most important rule of strong writing. All sentences have either active- or passive-voice verbs. Active voice emphasizes the performer of the action by making the performer the subject of the sentence. *Martha kissed John* expresses action. So does *A masked man with a shotgun robbed the Main Street Liquor Store.* Active verbs do something to someone or something. They are strong and direct, moving a sentence forward and making it easier and more interesting to read.

In contrast, passive-voice verbs put the focus on the receiver of the action, as in *John was kissed by Martha* and *The Main Street Liquor Store was robbed by a masked man with a shotgun.* Sentences cast in the passive voice are indirect and usually more wordy and less interesting than active-voice sentences. A string of sentences cast in the passive voice can bog down a reader.

Although in most cases sentences should be cast in the active voice, the passive voice sometimes is the better choice, such as when the emphasis should be on the receiver of the action. In these sentence types, the receiver of the action is more important or newsworthy than the agent of the action, as in the following example:

> Princess Diana's casket, wrapped in the maroon-and-gold Royal Standard, was carried out of Westminster Abbey by eight Welsh guards.[1]

1. "England's Rose at Rest," *The Star Tribune,* 7 Sept. 1997, A1.

move from the general to the specific, the same story may sound very different from its print sibling. A reporter for the Associated Press wrote the following for the print wire in 1996:

> The United States may be grossly underestimating the number of women who die due to pregnancy, the government reported today.[16]

The same story appeared on the AP broadcast wire from writer Russ Clarkson in this way:

> Pregnancy may be killing twice as many American women as anyone thought.[17]

As noted in Chapter 3, broadcast audiences have different needs than print audiences. Each can be served by summary leads, but usually the wording differs. In television writing, a summary lead is often spoken by the anchor, who introduces both the story and the reporter. For example, KARE, the NBC affiliate in the Twin Cities, uses this technique in its Mall of America parental escort story. Anchor Diana Pierce introduces the segment with a summary lead:

> Tonight's other big story takes us to the Mall of America. That's where a controversial new escort policy went into effect tonight. It forbids kids under age 16 from going to the Mall without an adult escort. But as Mall officials found out tonight, some rules are meant to be broken. KARE 11's Dennis Stauffer reports.[18]

The station then cuts to Stauffer at the mall, who continues the story by interviewing mall officials, security officers and kids, some of whom successfully sneaked in. The anchor provides the hard news of the curfew in the lead, and the reporter fills in the details. (For more details on the story, see pp. 142 and 144 in Chapter 6.)

The Multiple-Element Lead

A *multiple-element lead*, which can be a summary lead or another type of lead, works when a single theme would be too restrictive. In a multiple-element lead, the writer can use parallel structure to work more than one theme into the lead. Such leads are common in stories dealing with several actions taken by government bodies at one meeting. Here is an example:

> The Mayberry City Council Tuesday fired two department heads, established an administrative review board and authorized the mayor to begin searching for a new fire chief.

The reporter has concluded that all three of the actions taken by the city council are important enough to be in the lead. The reporter lists them in order of importance. Such a lead also serves to set up the rest of the story as details about each action are added. It is a multiple-element summary lead. The danger in such an approach, of course, is to try and do too much, cluttering the lead and confusing the reader.

In covering the Mall of America parental escort story, reporter Sally Apgar of the Star Tribune, the daily newspaper in Minneapolis, uses a multiple-element lead in a follow-up story on reaction to the proposed curfew:

> The Mall of America's consideration of a policy that would require youths under age 16 to be escorted by a parent or guardian after 6 p.m. on Fridays and Saturdays is drawing sharp criticism from community groups and teens but high praise from some storekeepers.[19]

The story then details the reaction from three groups — community organizations such as the Urban League, teen-agers and mall store owners — in the order in which they are presented in the lead.

Similarly, many leads in the stories about Target Stores' decision to stop selling cigarettes are of the multiple-element variety. This is so because of the company's justification of the move as an economic decision rather than an anti-smoking one. The Chicago Tribune's lead uses the multiple-element approach, and it adds another element:

> Claiming bottom-line economics rather than moral superiority, the Target Stores division of Dayton-Hudson Corp. broke ranks with other discount stores by declaring a smoke-free zone.[20]

The three-element lead promises to begin a story about economics and the morality of smoking, and includes a comparison to similar businesses.

Openings for Web pages tend to use multiple-element leads because of the desire to keep the Web reader moving from link to link. For example, among the most frequently visited Web sites is ESPN SportsZone, a product of the all-sports television and radio network. On the SportsZone Web page, the opening item is always one paragraph long and contains at least two links to stories, followed by a list of five or six links to related subjects. Here is how SportsZone handled its 1998 National Football League Super Bowl preview:

> Three current Broncos played in Denver's last Super Bowl loss in 1990, but don't expect them to freely share those memories. John Elway, Steve Atwater and Tyrone Braxton say they've < *blocked out their previous failure* > in the big game. Is Denver headed for another forgettable experience? The Zone seeks the answer in today's < *Stats Class* >.[21]

The opening provides links to the main story — "< blocked out their previous failure >" — and a statistical analysis of the game —"< Stats Class >." Five links to related stories follow the opening links. Note also the use of the present tense and the time element *today*. One advantage to writing for the Web is its immediacy, and sentence structure and word choice should reflect that.

The Suspense Lead

A *suspense lead* is the opposite of a summary lead. The writer manipulates facts and bits of information to leave the reader guessing about the main point of the story. The first couple of paragraphs set up the premise by providing clues; then a paragraph to follow gives the answer.

The suspense is often about why this person or that event is newsworthy. A suspense lead is problematic if the event is well known, and there is always a danger that the headline writer will give away the answer. But if the person or event is not famous, a suspense lead can draw readers into a story that they may not otherwise read. For example, here is a suspense lead:

SAVAGE, Minn. — Before Bo knew anything, before Air Jordan took flight, before George Herman Ruth became The Babe, there was Dan Patch.

At the turn of the century, Dan Patch became one of the most prominent sports figures of a generation by covering a mile in less than two minutes.

At this point, the reader knows Dan Patch is a successful sports figure and marketer from the early 1900s. The alert reader will figure out that Dan Patch is not human, because humans cannot run two-minute miles. The suspense is ended in the next three paragraphs:

It helped that he had four legs.

The mahogany-colored horse became the first commercial superstar of American sports. The pacer's name and likeness were lent to hundreds of products, including cars, beer, sheet music, kitchen knives and billiard cue chalk.

On June 20, 75 years after his death, the country's largest collection of Dan Patch memorabilia will be sold at the Meadowlands racetrack in East Rutherford, N.J. Experts value the 500-plus items at more than $100,000. The collection includes a workout sulky, a stove, a thermometer, postcards, posters and pails.[22]

Mystery solved. One flaw in a suspense lead is when the clues do not lead up to the answer, or, worse, when the writer forgets and leaves out the answer altogether.

It is a good idea not to hold the answer to a suspense opening too long, particularly if you are writing for the World Wide Web, where readers tend to have a short attention span. Here is how ScienceNews-Online uses one suspense opening:

The grandchildren who revamp Old MacDonald's Farm years from now may end up harvesting a product their forbearers ignored. If so, they can thank the first researchers to genetically engineer animals that concentrate a pharmaceutical product in urine.[23]

Here, the suspense and the answer to the "product" mystery are provided in the same paragraph — before the reader clicks on to another site.

Advertisements often attempt to add an element of suspense to capture the audience member's attention. But the suspense usually does not last long because of the space and time limitations in most ads, as well as the danger of losing an impatient customer. The Columbia Sportswear Company, makers of rugged outdoor clothing, featured a full-page color photograph of a camouflaged jacket, the type a waterfowl hunter would wear. The headline to the ad reads: "One possible explanation for Elvis' disappearance."[24] The ad was in Ducks Unlimited magazine, whose readers would be assumed to need camouflaged clothing and know who Elvis was (see Fig. 5-2).

The Character Lead

Many ad campaigns develop characters to get customers to identify with the products, from the Energizer bunny to the Maytag repairman. Print and broadcast stories use *character leads* to focus on individuals. They are usually found in feature stories, and they often contain rich, descriptive phrases about the person in the story. Character leads may contain information about a person's demeanor or appearance, or a description about some aspect of his or her life.

Newspaper reporter Terrie Claflin uses the following lead to begin one of a series of stories on Rachel, a baby girl born to a mother addicted to drugs:

> She is, in many ways, a china doll. Skin like snow, eyes like sky, a tiny body rigid and cool to the touch. Her cheeks are rosy, her face expressionless, unchanging. The world swirls in color and motion around her, yet she does not perceive it. For like a china doll, within her tiny head, behind those ice-blue eyes, Rachel has no brain.[25]

Character leads need not spend much time on descriptive phrases. Sometimes, briefly focusing on a real person can be used as an inlet into a larger issue. Articles that use examples of real people in the lead to explain complex stories are sometimes called "Dow-Jonesers" or Wall Street Journal-style stories, named after the newspaper company and newspaper that made them famous.

In the Target Stores example noted earlier, reporter Rekha Balu of Crain's Chicago Business uses a local woman to begin her story about pressure felt by other retail chains, including Illinois-based Walgreen's, to join Target in a cigarette sales ban.

One possible
explanation
for Elvis' disappearance.

 Our latest Omni-Quad™ parka is just the thing for people who don't want to be seen. Its powerful new Delta Marsh™ camouflage pattern has been refined to the point where it renders people practically invisible. Waterproof-breatheable Omni-Tech™ fabric, a zip-out insulating liner and roomy pockets make it the most versatile parka around. And just think, if it could make a man of Elvis' considerable dimensions disappear, imagine what it could do for you.

Columbia
Sportswear Company

For the dealer nearest you in the U.S. and Canada, call 1-800-MA BOYLE. WEB SITE: http://www.columbia.com

Figure 5-2

When Katherine Redd craves a nicotine fix, she goes to her corner Walgreen's and buys cigarettes.

When the 38-year-old data entry clerk wants to quit smoking, she scours the same store for nicotine patches and gums.

The retail paradox — a drugstore peddling a deadly and addictive product in one area and selling a potential cure in another — is not lost on Ms. Redd.[26]

Several stories on the Mall of America curfew also lead with an example of a person under age 16 not getting into the mall because of the new escort policy. Reporter Georgann Koelln of the St. Paul Pioneer Press adds a slight twist to that idea when she describes a mall employee who is nearly banned from work during the curfew:

"Are you kidding? I'm 18. I work here." It's 6:27 p.m. on Friday night. Petite, mini-skirted Galena Gregg, backpack slung behind her, is late for work at the Pacific Sunwear of California shop in the Mall of America, and she's being carded. She has no license, no state I.D. card.

To make matters worse, the whole country seems to be here witnessing the spectacle of this fresh fish caught in the net of the mall's new escort policy. . . .[27]

While some writers may disagree with the use of a direct quote to begin a story, the dilemma of the teen-ager is clearly and accurately portrayed by the writer. The story goes on to describe the incidents and experiences of the first night of the mall's curfew, including the crush of media people following teens around the mall.

The Scene-Setting Lead

Scene-setting leads are similar to character leads, except that instead of focusing on a person, the writer examines a place. This kind of lead is also rich in details, but the details are of the scenery — a park, a crash site, a place of business. People are not the main point of a scene-setting lead, but they may be included as part of the scenery.

Television stories, with their focus on visual images, commonly use scene-setting leads. The key is to let the pictures do the work. CNN reporter Larry Woods is an expert at matching words with visual images. In a series of stories called "Across America with Larry Woods," he crafts descriptive, informative text to match majestic pictures of the mountains, the sea and the desert. In a story on fabled U.S. Highway 66, under a backdrop of Arizona scenery and a ribbon of highway, Woods intones in his distinctive voice:

> In the high desert country of western Arizona, they are try-ing to bring back the past . . . trying desperately to bring back to life what John Steinbeck called in 1939 'the mother road of America . . .' Historic Route 66.[28]

A scene-setting lead may also include elements outside the physical environment of the place itself, helping to establish a mood. Such is the case with Bill Barich's story about a boxing club for Sports Illus-trated. He begins like this:

> Professional boxing has always thrived in an atmosphere of greed, lar-ceny, poverty and casual violence, so it has always been at home in Philadel-phia, a tough town with an attitude, where the mob still dumps a few bodies into the Delaware River every year. Open the Yellow Pages in Philly, and the first thing you notice are ads for a pair of ambulance chasers, who offer a seductive menu of potentially actionable mishaps — bus accident, dog bite, slip-and-fall. Unemployment is high, scams mutate and multiply, bookies proliferate, and hard drugs are easy to find. Crack cocaine has turned cer-tain blocks into piles of rubble as bombed out as any in Belfast or Beirut, and the young men who live there are sometimes desperate enough to put on the gloves and aim themselves toward the Legendary Blue Horizon in hopes of escaping.
>
> There isn't another sports arena in the country remotely like the Blue Horizon. It's the sort of raw and smoky cavern that George Bellows painted early this century, a throwback to the era of straw hats, stogies and dime beers. Only 1,500 fans can be crammed inside for an event, but the crowd compensates for its lack of size with its animal howling. . . .[29]

Outside of the reference to the painter, the first few hundred words of the story go by without any mention of an actual person, much less a quotation. The story is about the building, and the people inside are a part of that story — as are the city, the bombed-out piles of rubble, the river and the drugs.

The Narrative Lead

Character and scene-setting leads are part of a larger style of writing called narrative. The *narrative lead* may set the scene or describe a character, or it may use an anecdote to illustrate a larger point. The key to this type of lead is the inclusion of a nut paragraph, which is, in essence, the *who-what-when-where* section that tells the reader why the story is important. The nut paragraph should reply to the statement "Here's why you should care about this story."

Sometimes a narrative lead is chronological, in a story-telling sense. Mitch Albom of the Detroit Free Press describes the scene at a press conference at the high school of a prep basketball star in chronological order as a way of beginning his comparison of the emphasis placed on sports versus academics in America:

The little chocolate doughnuts were in a box, next to the coffee urn. Normally, high schools don't provide food for their assemblies, but today was special, all these TV crews, radio people, sports writers. A table was arranged near the front of the room, and a reporter set down a microphone, alongside a dozen others. "Testing 1-2 . . . testing 1-2," he said.

Suddenly, the whole room seemed to shift. The guest of honor had arrived. He didn't enter first. He was preceded by an entourage of friends, coaches, his grandmother, his aunt, his baby brother, more friends, more coaches and his girlfriend, whom he identified later as "my girlfriend." She wore a black dress and jewelry and had her hair pinned up, as if going to the prom, even though it was mid-afternoon and math classes were in progress upstairs.[30]

After some more description of the scene, Albom then goes upstairs to the math class and finds another kid with a scholarship — an academic scholarship — and compares the two youngsters. Narrative writing uses elements of fictional writing, including setting the scene, telling a story, adding suspense and exploring characters (see Chapter 8).

Avoiding Overused Leads

University of New Hampshire journalism professor Jane Harrigan offers a list of overused leads that beginning and experienced writers alike sometimes fall back on.[31] In addition to her suggestions, several other ineffective leads should be avoided.

Trite Reference Book Leads. Among the most overused, dull lead forms is the quick trip to the nearest reference source on the writer's desk, particularly the calendar, almanac and dictionary. If the writer sees the calendar first and a holiday is approaching, watch out! Needless references to Thanksgiving, Valentine's Day and April Fool's Day are especially common in calendar leads: "Yesterday was Thanksgiving, but Jane Smith was no turkey as she was elected president of the school board." While that may be an extreme example, consider that references to the former National Basketball Association player Darnell Valentine showed up in countless sports pages on Valentine's Day as a calendar lead.

The almanac also has supplied many a stuck writer with the grease to wriggle free. The almanac is used to verify dates in history, in the writer's attempt to put events in context. But the almanac's data may have nothing to do with the events at hand: "When Jim Johnson was born, gas was 27 cents a gallon, bread was 15 cents a loaf, Lyndon Johnson was in the White House and man had not yet reached the moon." Unless Johnson is a gas station attendant, a baker, a relative of a dead president or an astronaut, this lead makes no sense. Only about a zillion people fit that description. Yet writers sometimes try to force the issue, particularly when their subject is unusually old or very young. This type of lead works only when a direct relationship exists between the person and the historic events.

Dictionary leads are not as common as they once were, perhaps because computer spell-checkers have rid the writer's desk of the book version. However, we still see leads of this type, as in the following example: "According to Webster's, a frog is defined as a green, often slimy reptile that lives in swampy areas." Uh huh. The readers already know that. Tell them something they do not know. (One particularly odious offspring of this lead is the pictionary lead: "If you looked up courage in the dictionary, you would find Neil Armstrong's picture beside it.")

Powerless Declarative Leads. Declarative leads try to shake the reader or listener to attention with a seemingly powerful word or short phrase. Sometimes the phrase is not a true fact: "There's a wolf in the woods! Wolf! Run! That's the word from the boy who cried it Monday. . . ." The danger is that the person who stops reading after the declarative lead may panic and start running from the wolf.

At other times, the phrase *That's the word* may be supplanted by *It's official* when the incident becomes fact. The *It's official* lead may appear after a writer doing a series of stories on an unresolved issue has to write a final story about the resolution, such as a government authority's decision. Among its other problems, this lead gives too much credence to the persons or groups making the decisions.

Another type of overused declarative lead is the mystery *It* lead: "It is round. It is red. It is edible. It is a tomato, and the fall crop looks good for central valley farmers this year." Riddles — especially those that insult the intelligence of the audience — are best left for other venues. Besides, the headline writer always gives it away: "Tomato Crop Looking Rosy This Fall."

One-word leads are rarely effective, unless the word is especially powerful or striking. Too many stories about roller coasters, for example, start out with a word like "Yahoo!" The writer then spends the entire second paragraph explaining the one-word lead, when that paragraph could have stood alone as the lead.

Weird Linkages. The uncommonly common lead always links two or more weird items: "What did Boris Karloff, Jane Austen and Richard Nixon have in common? Ring around the collar." Besides failing the "Who cares?" test of a lead, this type of lead is usually foiled by the headline or introduction. A variation on the uncommon is the atypical: "At first glance, Jenny Smith looks like a typical college student. But beneath her appearance, she's really Batgirl." Very few news stories are written about people who are "typical." Assuming, then, that the subject is in some way exceptional, why begin a lead with, "Most politicians love to glad-hand, and John Doe is no exception. . . ."? If John Doe is not an exception, why write about him? Find someone who is exceptional and do a story on that person.

Also in this category of overused leads are the so-called Dangerfields, named after the comedian with the famous signature line, "I don't get no respect." Here, the writer leads with an unfounded claim about a profession, such as nurses, firefighters, lawyers or cab drivers, "never getting any respect": "Night-shift nurses get no respect, and Nellie Nightengale is no exception. . . ." Often these leads refer to work-related groups and their claimed lack of respect, but other types of people — perhaps identified by gender, race, ethnicity, education, social class — and other stereotyped ideas about them would also fit into this category.

Irrelevant Chair-Leaners. The chair-leaner lead has some characteristics of the scene-setting and character leads, except that the person is described as leaning forward or backward in a chair: "The man leaned forward in his chair and gazed at a picture of his dead wife." Setting the scene is important, but which direction the chair is tilted is not significant.

An offshoot of the chair-leaner is the question-ponderer: "The woman leaned back in her chair and pondered the question." Here, the writer is (1) attempting to set the scene, and (2) congratulating him- or herself on posing a question that the subject has to ponder before responding.

Uninformative Question Leads. Question leads appear in print more often than they should. Although they are sometimes effective, in most cases a question lead results in the reader's answering "No" or "I don't care," at which point the reader bails out. In the worst case, the writer forgets to answer the question and the reader is left wondering. It is relatively easy to turn a question lead into a short statement that provides more information for the reader. For example, the question lead "Wonder how your trees will get trimmed during the strike?" could be easily revised to read this way: "Some tree-trimmers are set to cross picket lines tonight."

THE NEXT STEP: WRITING BASIC STORIES

A well-written summary lead is a condensed version of the story, providing basic details about *who, what, when* and *where*. Take a look at any daily newspaper's sports section, for example, to see how the results of a college football or basketball game are reported. Typically, the entire story is condensed in a one-sentence opening paragraph:

> ST. PAUL, Minn. — Senior guard Karnell James scored 47 points, two shy of the conference record, to lead St. Thomas to an 87-85 victory over Carleton Saturday night at O'Shaughnessy Arena.

The lead summarizes the basic story, telling *who* won, *what* the score was, *where* the game was played and *when* it was played. In fact, most newspapers in Minnesota use only one paragraph to describe a game.

Length is one way to define a basic story. Most stories carried on the Associated Press newswire are less than 400 words, and news releases tend to occupy fewer than two double-spaced pages. Broadcast stories can be one or two sentences in length, particularly in the case of headline news on the radio. The majority of advertisements face boundaries of space or time that force them to be brief as well.

Calling a piece of writing a basic story can also refer to the subject matter dealt with by the writer. For our purposes, stories dealing with a single issue, incident or person can be considered a basic story, although their length might be as much as 1,000 words or more than two minutes of broadcast time. In Chapter 6, then, we look at how to write basic stories.

NOTES

1. California Milk Processor Board ad.
2. Jamie Hanrahan, "Milk Mustaches for Everybody," *The Los Angeles Daily News*, 3 Feb. 1997, L5.
3. Rene J. Cappon, *The Word* (New York: Associated Press, 1982), 31.
4. Archibald MacLeish, quoted in Eric Barnouw, *The Golden Web* (New York: Oxford University Press, 1968), 151.
5. Franklin D. Roosevelt, quoted in Michael Emery and Edwin Emery, *The Press and America*, 8th ed. (Englewood Cliffs, N.J.: Prentice Hall, 1996), 346.
6. Allen H. Center and Patrick Johnson, *Public Relations Practices* (Englewood Cliffs, N.J.: Prentice Hall, 1995), 275.
7. "Target Issues Media Statement on Cigarette Sales," news release, PR Newswire, 28 Aug. 1996, 1.
8. Brad Woodard, "Piper Crash," KARE 10 p.m. news, 3 June 1996.
9. Ring Lardner, quoted in Peter Andrews, "The Art of Sportswriting," *Columbia Journalism Review*, May–June 1987, 27.
10. Christi Parsons and Gary Marx, "Fiery Crash on Runway Claims 13," *The Chicago Tribune*, 20 Nov. 1996, 1.
11. Bill Brubaker, "Ex-Commissioner Rozelle, Architect of NFL, Is Dead," *The Washington Post*, 7 Dec. 1996, A1.
12. "Mall of America to Implement Parental Escort Policy," news release, Mall of America, 4 Sept. 1996, 1.
13. "Eating Lake Ontario Fish Linked to Shorter Menstrual Cycles; Consumption May Delay Pregnancy, UB Researchers Find," UB News, accessed 2 Dec. 1997 at http://www.buffalo.edu.
14. Jessica Lee, "Gingrich Wins Close Vote," *USA Today*, 8 Jan. 1997, 1.
15. Bruce Benidt, "IDS letter," Mona, Meyer & McGrath files, Minneapolis, Minn., 5 Oct. 1989, 1.

16. Reported in Brad Kalbfeld, ed., "Closed Circuit," Associated Press memorandum, 2 Aug. 1996, 2.
17. Ibid.
18. Diana Pierce, "Escort Policy," KARE 10 p.m. news, 4 Oct. 1996.
19. Sally Apgar, "Megamall's Plans for Required Escorts Praised, Assailed," *The Star Tribune*, 21 June 1996, 1B.
20. George Gunset, "Aiming at Bottom Line?: Target Moves to Stub Out Cigarette Sales," *The Chicago Tribune*, 1 Sept. 1996, 2.
21. "Orange Crushed: Painful Memories," ESPN SportsZone, accessed 22 Jan. 1998 at http://www.ESPN.SportsZone.com.
22. Mark Neuzil, "A Town's Heritage Put up for Sale," Associated Press, 18 May 1992.
23. "Future Farmers May Collect Urine, Not Milk," ScienceNewsOnline, accessed 10 Jan. 1998 at http://www.sciencenews.org.
24. Columbia Sportswear Co. advertisement, *Ducks Unlimited*, Jan.–Feb. 1997, 97.
25. Terrie Claflin, "Mother's Drug Habit Turns Baby into Victim," *Medford Mail Tribune*, 19 July 1989. Reprinted in *Best Newspaper Writing, 1990*, ed. Don Fry and Karen Brown (St. Petersburg, Fla.: Poynter Institute, 1990), 2.
26. Rekha Balu, "Could Walgreen's Kick Tobacco Habit?" *Crain's Chicago Business*, 14 Oct. 1996, 1.
27. Georgann Koelln, "National Media Circus Hits Mall of America," *St. Paul Pioneer Press*, 5 Oct. 1996, 1F.
28. Larry Woods, "Across America with Larry Woods" videotape (Atlanta: Turner Multimedia, 1992).
29. Bill Barich, "Singin' the Blue," *Sports Illustrated*, 5 Dec. 1996, 73.
30. Mitch Albom, "Why Do We Focus on Body over Mind in High School?" *Detroit Free Press*, 11 April 1995. Reprinted in *Best Newspaper Writing, 1996*, ed. Christopher Scanlan (St. Petersburg, Fla.: Poynter Institute/Bonus Books, 1996), 302.
31. Jane Harrigan, "Editors Should Stamp out These 10 Overused Leads," *ASNE Bulletin*, Sept.–Oct. 1987, 20–21.

6
WRITING BASIC STORIES

... Thirty years ago my older brother, who was 10 years old at the time, was trying to get a report on birds written that he'd had three months to write ... he was at the kitchen table close to tears, surrounded by binder paper and pencils and unopened books on birds, immobilized by the hugeness of the task ahead. Then my father sat down beside him, put his arm around my brother's shoulder, and said, "Bird by bird, buddy. Just take it bird by bird."

— Anne Lamott, "Bird by Bird"[1]

Building a good story is like crafting an exquisite piece of furniture. They both demand good tools and materials. For writers, the tools are an understanding and mastery of grammar, including punctuation, parts of speech and sentence construction. But the difference between knocking out a story that resembles a plywood dog bed or crafting one that shines like an oak veneer coffee table is in the quality of the raw materials. In basic story writing, the raw materials are facts, details, descriptions and definitions.

Advertising executive Sheldon Clay, one-half of a creative team on his ad agency's Harley-Davidson account, tells about his experience looking for a few good details and descriptions while standing in front of a gas station just outside of Talladega, Alabama. As the rays from the early morning sun bounce off the silver caps of the gas pumps, a half-dozen riders sit astride their Harleys. Clay is on a nine-day photo expedition from Daytona Beach to Milwaukee, the home of the Harley. He and his advertising crew want pictures of the bikes and their riders as they wheel down interstate freeways and backwoods highways. The photos will be featured in a glossy brochure for the motorcycle's dealers. As Clay looks over the riders and the bikes with their gleaming polished chrome and high-gloss paint, he notices a middle-aged man standing next to him. The man, who is also looking

"Love must be what you feel when you like something as much as you like your Harley-Davidson."
—*Overheard among bikers.*

There's devotion, and then there's whatever you call what gets inside the heart and mind of the Harley-Davidson rider. Mom should be so well-loved as the average Harley-Davidson motorcycle. The United States Marine Corps should inspire such loyalty.

You will see them in the wind, riding Harleys® from all eras.

The preacher who roams the country on a Harley-Davidson Sportster® with an angel painted on the fuel tank.

The man who spent 13 years and wore out three cars tracking down every single original part needed to build a 1958 Panhead because he came across one in a photo one day.

The retired couple who've shown up at the Black Hills Motorcycle Rally in Sturgis, South Dakota on their vintage Harley-Davidson police motorcycle every August for 43 years. (About 57,000 miles just going there and back.)

The rider who got on his friend's XLCH one day in 1961 and ha[s] it halfway around [the] world in the thirty-[years] since and still show[s] signs of slowing, or trading it out for a ne[w]

There are thous[ands] thousands out ther[e] so far gone their re[...] can only shake the[ir] heads and sigh. Bri[...] bankers, secretari[es]

Chpt 16: **THE HOPE**

We care about you. Sign up for a Motorcycle Safety Foundation rider course (for info call 1-800-447-4700). Ride with your headlight on and watch out for the other person. Always w[...]

Figure 6-1

at the bikes, has a beard and long, brown hair that covers the collar of his work shirt, which is tucked into blue jeans. "I'm pretty shy," Clay recalls. "But I've got enough courage to go up and say 'Hi,' and then I listen." As Clay listens, the man tells him that he is a preacher and that

he organizes motorcycle trips for fellow believers. He also tells Clay that he owns and rides a 1962 Harley-Davidson Sportster, with an en- gine he rebuilt. The motorcycle is blue, he says, and it has an angel painted on the gas tank. "I always have a piece of paper in my pocket," Clay recalls, "so I can jot down a few details. I wrote that one down."[2]

The angel ended up in a 1997 Harley-Davidson ad series (Fig. 6-1) entitled "The Book of Harley-Davidson," in chapter 16, "The Hope-lessly Addicted." The ad begins in this way:

> *"Love must be what you feel when you like something as much as you like your Harley-Davidson." — Overheard among bikers.*
> There's devotion, and then there's whatever you call what gets inside the heart and mind of the Harley-Davidson rider. Mom should be so well-loved as the average Harley-Davidson motorcycle. The United States Marine Corps should inspire such loyalty.
> You will see them in the wind, riding Harleys from all eras.
> The preacher who roams the country on a Harley-Davidson Sportster with an angel painted on the fuel tank.[3]

The ad goes on to describe a retired couple who have shown up on their Harley at the Black Hills Motorcycle Rally in Sturgis, S.D., every August for 43 years. It also features a rider who borrowed a friend's Harley in 1961 and, in the 36 years that followed, rode it halfway around the world. Overall, the ad is successful because its writers include a wealth of facts and details.

Telling a good basic story depends on good details. Television producer Pat Weiland and reporter Cindy Hillger fill their 1989 follow-up story with details about the rash of car-train collisions from the train engineer's perspective:

> Milt Rae holds the throttle on 7,000 tons of steel traveling at 50 miles an hour. His train is a mile long. And it would take a full mile for him to stop the train.
> (The engineer talks.)
> For every pound of car, there are 6,000 pounds of train bearing down on it. The impact would be the same as if you drove your car over a 12-ounce pop can.[4]

Weiland and Hillger provide specifics for their train story. They resist adjectives and stick with the facts. The train is not only big; it is "7,000 tons of steel." It is not traveling at a fast or slow speed, but at "50 miles an hour." It is not just a long train; it is "a mile long." And to enhance the viewer's image of the impact between a car and train, the writers add a head-turning analogy: It "would be the same as if you drove your car over a 12-ounce pop can."

Sometimes, compelling details are used simply as an example. In the following passage, the 3M Company public relations writer wants to demonstrate the tenacious quality of the company's Post-It notes on the occasion of their 10th birthday. She points out that Post-It notes are one of the five top-selling office supply products in the United States. She adds that they were invented by Arthur Fry, who wanted a bookmark that would not fall out of his hymnal while he sang in the church choir. Finally, the writer closes with an example that forcefully makes her point:

> Over the decade, thousands of people have written fan let-
> ters to 3M expressing their enthusiasm for Post-It notes. One of
> the most extraordinary came in October 1989, from a survivor of
> Hurricane Hugo in Charleston, South Carolina. Bruce Brakefield
> returned to his home after it had been battered by 140-mph
> winds for three days. Incredibly, the Post-It note he and his wife
> used to warn visitors when their baby was sleeping was still stuck
> on his front door. Wrote Brakefield, "This little piece of paper
> withstood Hurricane Hugo whereas eight oak trees in my yard
> did not."[5]

FINDING A FOCUS FOR THE STORY

Facts, details, examples — these are the raw materials for building a basic story. The writer also needs a blueprint, a construction plan for crafting the story. Part of that plan involves finding the theme or focus of the story. The writer asks, "What is this story *really* about?" The answer ought to provide more than a beginning; it should also provide some of the threads that will tie the story together.

Those threads frequently fit the more traditional definition of news. The writer thus looks for an element that is timely, a person who is prominent, an event that is nearby, an occurrence that is differ-ent, a finding that affects may people's lives or a moment that attracts national attention. The writer selects the threads that will best convey the focus of the story.

Tell the Story with Words

Alice McQuillan of The New York Daily News crafted a crisp 10-paragraph story in 1997, about the rescue of a Brooklyn boy who almost drowned. Her focus in the story is the rescue. It is a simple blueprint that eases her from a summary lead to paragraphs that follow. Here is the lead:

> A 14-year-old Brooklyn youth spent 20 harrowing minutes in the frigid waters of a pond yesterday after falling through the ice.[6]

Next, McQuillan gives more information, such as the victim's name, the reason he was on the ice and how far he was from shore:

> Anthony Flax was taking a shortcut across the pond when the ice began to crack beneath him, leaving him stranded 30 feet from shore.
>
> "I thought it was safe because it was so cold," said Flax, who said he will never again walk across the pond near his Plum Beach home.[7]

Then she describes the rescue, using good quotes, short sentences, colorful details and a simple chronology:

> It took about a dozen cops, using a rope tied to a long stick, to pull him from the water. "It looked like he was about to go under any second," said Officer Brian Billy of the 63rd Precinct.
>
> "He was going lower and lower. He was saying 'Help! Help! I'm freezing.' We kept telling him, be calm."
>
> Luckily for Flax, a passerby heard his frantic screams when the ice cracked and he plunged into the weed-choked pond at about 11 a.m. yesterday.
>
> The temperature then was 12 degrees, and Flax became submerged to his chest.
>
> "I was out there in the water for five minutes and on the ice for a good 15 minutes, half in and half out," said Flax, who is an eighth-grader at Shell Bank Junior High School.
>
> Officer Vincent Varvaro of the housing bureau broke his ankle pulling Flax to safety.[8]

The New York Daily News writer uses 340 words to tell her story for the print medium.

Online writers are even more economical in telling stories for their audience on the World Wide Web. The idea is to tell the story quickly and succinctly, and to let readers click on to other stories or Web sites at will. At the CNN/Time magazine All Politics site, a story posted in 1997 began with this 29-word lead about U.S. Attorney General Janet Reno's decision:

> Attorney General Janet Reno has decided to ask a three-judge panel for an extension in the probe of President Clinton's fund-raising until Dec. 3, CNN has learned.[9]

The writer then answers the readers' question: Why?

> Sources close to the investigation say the FBI asked Reno for more time to clean up technical issues related to the probe.[10]

For readers unaware of the status of the probe, the writer gives an update:

> As late as Friday, Justice Department prosecutors were thinking about shutting down the investigation, after finding no specific, credible evidence that any crime took place.[11]

Next, the writer deals with why the FBI has yet to sign off on the investigation, easing the transition between paragraphs with the word *but*:

> But the FBI made it clear over the weekend that it had not completed its investigation. The bureau wants to re-interview some witnesses, sources say.
> Justice officials also wanted additional time to make sure they got the matter right.[12]

The following paragraph gives some needed background about the probe, but note how the writer keeps it brief:

> The FBI wants to know if Clinton conversations led to soft-money campaign contributions being shifted to so-called hard-money accounts.[13]

In addition to this brief background, however, several links to sites with more information are positioned next to the story's main text. Readers can click on links to access stories about Clinton's White House tapes and to visit a related U.S. Justice Department site.[14] In much online writing, readers decide how much background material they want or need to read.

Tell the Story with Pictures

The attention span of information seekers is of particular concern to advertising writers and television news producers, who have only seconds to attract a customer's or viewer's eye. They use pictures to help make the point.

The ad writers for Imation, an information and imaging company spun off from 3M, use a black-and-white picture, a two-word headline and a 12-word deck to make their point in the ad shown in Figure 6-2. Although the ad copy uses only 75 words, the focus of the ad is sharp, its message is crisp, and its effect is memorable. The writers tell the whole story: 3M is launching a new company, with the same creative qualities of its parent. Note also that the copy makes use of sentence fragments. Although beginning media writers should generally avoid experimenting with fragments, in the hands of a veteran they can help move a story along.

Television news writers also want to move their stories along. Most standard TV news stories, called *voice-overs*, run 30 seconds or less. Here is an example from a 1996 KARE newscast:

> Television and film star Woody Harrelson is free on bond tonight, after being arrested for cultivating marijuana.
>
> His arrest was part of a staged event.
>
> Harrelson planted marijuana seeds in front of a sheriff in Kentucky.
>
> He did it to call for new laws differentiating between marijuana and its less potent cousin, hemp.[15]

The story's focus is the arrest of Woody Harrelson, and the writer adds a few details such as those telling *where* and *why*. There is little time for more. The idea is to get at it, get with it and get it done. The longest paragraph has 17 words; the shortest sentence has eight words.

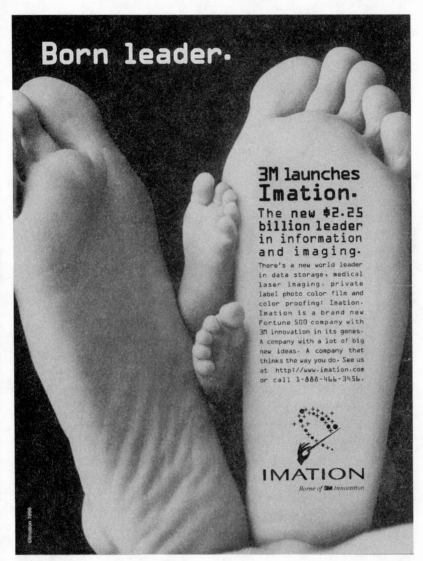

Figure 6-2

The simple voice-over is a standard element in most television newscasts, and producers write dozens of them each day. However, in some cases reporters tell stories that extend for two or three minutes of videotape, rather than 30 seconds of an anchor reading over pictures. One example comes from Ken Speake, a master storyteller who writes in a conversational style while keeping his eye glued to the focus of his story. In 1994, Speake traveled to a Mississippi River valley to meet an 87-year-old man who had become somewhat of a legend among his neighbors because of the gladioli he grew. The man told Speake that the flowers were his life's work. The reporter richly conveys that sentiment in his script:

> Look at 'em, tall on their stately stalks, frilly florets of absolutely amazing colors. This is Carl Fischer's gladiolus field. A field that many folks might watch and enjoy, Carl Fischer worships.
>
> It's not the gladiolus that he worships, though he's worked with 'em for more than 50 years now.
>
> It's more that his glads provide a kind of church, in the midst of which Carl Fischer feels a perfect peace. (*Here there is a sound bite from Fischer, in which he says he feels in harmony with nature, in harmony with God.*)[16]

The story, which focuses on the flowers and the old man's feelings toward them, took up three minutes and two seconds of air time. It included a few pictures of Fischer, his gladioli and his gladiolus field, and a few sound bites from Fischer. The result is a compelling and colorful story that provides a focus and makes a point.

ORGANIZING THE STORY

While the focus of a basic story may be obvious, figuring out how to organize a story may be more challenging. Writers of longer stories frequently make an outline, a blueprint identifying the parts of the story and, more importantly, how they fit together. The writers may be sorting through a hundred pages of notes, hours of transcribed interviews and dozens of quotes from background articles. An outline usually starts with the lead, or most important point, highlights those facts and examples that support or elaborate on the lead, identifies any cause-and-effect relationships, seeks transitions from one part of the story to the next and reveals an anecdote or quotation to summarize the main points.

With a basic story, the process is similar, but simpler. The task is to tell the story with a look-at-this beginning, a move-it-along middle and a tie-it-together ending. Examples from a story covered in October and November 1995 will help show how the print and broadcast media accomplish these goals using different strategies.

The newsworthy events of the story we will examine occurred during the halftime of a "Monday Night Football" game between the Minnesota Vikings and Chicago Bears. The head of the Vikings' security force scuffled with a Twin Cities television station photographer, and the episode was caught on videotape by a freelance photographer. Local newspapers and television stations reported on the incident in the following days. The focus of their stories was the same: The photographer claimed he had been roughed up and thrown to the ground by the guard. But how the print and broadcast reporters organized their stories differed in important ways.

The Star Tribune's account is a straightforward, just-the-facts kind of story. It begins with this lead:

A freelance photographer's video shows a Channel 9 photographer being thrown to the ground by the head of Vikings security during the Vikings-Bears game on Monday night. Photographer Dan Metcalfe had been arguing with security about where he was standing and was taken away by the head of Vikings security and a Minneapolis police sergeant.[17]

At this point, newspaper writer Anne O'Connor knows she is obligated to tell readers what happened. She does so by describing what her print audience cannot see — the episode that was captured by chance on videotape.

The video shows Metcalfe being thrown to the ground by Steve Rollins, the director of security for the team. Rollins continues to push Metcalfe and pulls him out of the tunnel with the help of Minneapolis Police Sgt. Don Banham.[18]

Several paragraphs later, O'Connor continues her description of the video:

But the video clearly shows Rollins grabbing Metcalfe by the back and hurling him to the ground. Metcalfe's camera fell to the side. As Metcalfe is being escorted out with the sergeant and Rollins, he is thrown to the ground again.[19]

Covering the same story for a KARE broadcast, television reporter David Wildermuth does not have to write with the same level of detail as O'Connor does in describing the incident for a print audience. His viewers can *see* what happened in the video. Therefore, he organizes his story differently, in a way that allows him to be more conversational, and perhaps more suspenseful. Wildermuth begins his videotape story with pictures of the game and the teams leaving the field at halftime, and then he writes:

> Even if you did stay on the couch for the halftime show, you missed the real action. It wasn't on the field; it was in the Vikings' tunnel. For the highlights we go to the videotape.
>
> (Here, the sound comes up full and the photographer is heard shouting, "C'mon, don't touch me.")
>
> That's Channel Nine Photographer Dan Metcalfe. The man who put him on the ground — the head of Vikings security Steve Rollins.
>
> (More ambient, or natural, sound from the scuffle.)
>
> Off-duty Minneapolis policeman Don Banham joins in and carts Metcalfe off in a neck hold, the photographer stumbling and complaining the whole way up the tunnel.[20]

Note that Wildermuth's version is much chattier than his newspaper counterpart's: "Even if you did stay on the couch for the halftime show, you missed the real action." His sentences are generally shorter: "That's Channel Nine Photographer Dan Metcalfe. The man who put him on the ground — the head of Vikings security Steve Rollins." Wildermuth does not waste words, and the words he chooses are strong nouns and action verbs: ". . . policeman Don Banham *joins* in and *carts* Metcalfe off . . . the photographer *stumbling* and *complaining* the whole way. . . ." Wildermuth writes this way because the story will be read aloud.

There are still other differences between the print and broadcast versions of this story. Consider, for example, how each version ends. O'Connor's ending describes the reactions of the Vikings and the organization's head of security:

> After Monday's incident, the organization released a short statement basically confirming Metcalfe's arrest (a disorderly conduct charge against the

photographer was later dismissed) after "repeated requests" to move.

Rollins declined to comment except to say that he "will have an appropriate response."

"I just don't have a response for you right now," he said.[21]

This ending is typical of print stories using the inverted pyramid structure. The most important facts come first; then the story winds down to the ending, where the least important facts are given. In the preceding example, the quotes do not add much information to the story and are appropriately put in the ending. The inverted pyramid approach, then, does not require the writer to provide a conclusion to the story.

Broadcast news stories, in contrast, are told aloud in a storytelling fashion. The writer is required to satisfy listeners' and viewers' expectations that the story will have a traditional conclusion. Notice how Wildermuth wraps up his story with a live stand-up (on camera) from the newsroom, bringing it to a satisfying conclusion:

> Don Banham, the off-duty policeman, told me tonight that Metcalfe was verbally abusive and "acting like a jerk" before the scuffle. A number of season ticketholders who watched things unfold called us tonight to tell us the incident was unprovoked, and the guard was overzealous and out of line.
>
> Metcalfe was handcuffed and charged with disorderly conduct. He says he isn't thinking of suing tonight. He'd just like this to go away.[22]

Although both writers have the same facts, each organizes the story differently, according to the needs of the broadcast or print medium. Public relations writers also have their own way of organizing press releases. In the case of the Vikings, the PR writers' task is not to restate the facts of the story but to let the community and fans know that the team is concerned about the incident. As we saw earlier, the first response is vague, acknowledging only that the photographer had been arrested after repeated requests to move. The next response, quoted in The Star Tribune, clearly conveys the team's position on the issues raised by the incident:

WRITING TIPS

USING THE INVERTED PYRAMID STRUCTURE

The inverted pyramid structure is most useful to print media writers without a lot of time or space. The topsy-turvy triangle demands that such writers get to the main point of the story quickly, answering as many of the *who, what, where, when* and *why* questions as possible in the lead. From there, writers add less important facts or anecdotes, material that can be trimmed by a copy editor seeking to eliminate column inches. What the style lacks in subtlety, it makes up for in efficiency. Consider the following example:

> SANTA ANA, Calif. (AP) — John J. Famalaro should be executed for the brutal sex killing of a waitress whose body was stashed three years in an Arizona freezer after the slaying, a jury decided Wednesday.[1]

In just 30 words, the writer tells *who* (John Famalaro), *what* (should be executed), *why* (for a brutal sex killing, the jury decided) and *when* (the verdict came on Wednesday).

The next paragraph lets readers know how old the man is, who he killed, how old the victim was and when he killed her.

> It was the same jury that convicted the 40-year-old former handyman last month of murder and special circumstances of sodomy and kidnapping in the 1991 death of Denise Huber, 23, of Newport Beach.[2]

By the time readers reach the last two of the story's six paragraphs, they have been given supporting details: The victim was struck 31 times, with a nail puller, in a Laguna Beach warehouse.

---

1. Associated Press, "Jury Urges Death for Con in Brutal Sex Killing," *The Arizona Republic*, 19 June 1997, A2.
2. Ibid.

"Regardless of the provocation, there is no justification for the type of force used in the incident," said Vikings president Roger Headrick. "Appropriate disciplinary action is being taken with those involved . . . to ensure that there is no repetition in the future.

"We regret the occurrence of this incident and do not condone and will not accept this type of behavior."[23]

In many ways, good public relations releases are similar to news stories, using crisp, clear writing to make a point and to capture the reader's attention. Unlike news stories, however, PR releases aim to create a positive image of the client; for example, as responsive to criticism, concerned about customers and eager to correct mistakes. Given the mild public reaction to the photographer incident, we might conclude that the Vikings' PR response helped to calm potential critics by portraying the organization and its strong stance on the issue in a positive light.

WRITING THE STORY

Bring the Pieces Together

Once the media writer has a focus for the story and a plan for organizing the facts coherently, the next task is to stitch the pieces together — sentence by sentence, paragraph by paragraph, page by page. A good outline or design can help make this task less daunting.

As noted in Chapter 5, a basic print news story begins with the lead. Here, the writer attempts to summarize the most important facts of the story. A lead therefore contains a burst of information, and writing it is similar to recounting an event or experience to someone who was not present when it occurred. In the paragraph following the lead, the writer adds essential details, maybe even making a sales pitch for the rest of the story to keep readers interested. In each succeeding paragraph, still more details are introduced in declining order of importance, nudging the reader along a clearly marked path.

Whenever possible, the story is told chronologically, relating the events in the order in which they occurred. The writer also seizes opportunities to highlight good quotes or sound bites in the story and to

WRITING TIPS

USING THE NARRATIVE, CHRONOLOGICAL AND HOURGLASS STRUCTURES

As noted in the Writing Tips box on page 138, the inverted pyramid structure provides a fast and efficient way for organizing a story that is limited by time or space. But it is not necessarily the most interesting or subtle method. Experienced writers and storytellers may prefer a less structured, more imaginative method for piecing a story together, particularly when they are allowed the space and time to do so. They may use a narrative, chronological or hourglass structure.

Narrative form is used to describe scenes and people, often incorporating dialogue between the people in a story.

Chronological structure is often used to tell a story within a story. As the term implies, the story begins with an interesting moment and follows the action as it unfolds. A chronology is particularly effective in stories about a crime or an accident, for instance.

Hourglass structure combines the inverted pyramid, narrative and chronological forms. According to writing coach Roy Peter Clark:

> The top of the hourglass looks like the old inverted pyramid, but is shorter in duration — perhaps four or five paragraphs. So we learn that a man shot a police officer in the leg, ran into a house, held a boy hostage for eight hours, surrendered without harming the boy, and was finally arrested. What follows is a transition, called the *turn*. "Police and witnesses gave the following account of the dramatic incident." What follows is a retelling of events in chronological order, with many more details than a standard story would allow.
>
> Readers now have a choice. They can read the top and quit, or if interested can linger down in the story.[1]

1. Roy Peter Clark, "Two Ways to Read, Three Ways to Write," *Workbench*, Winter 1998, 12.

give emphasis to issues of human interest. This is the general blueprint that Star Tribune reporters Dan Wascoe Jr. and Sally Apgar follow in a 1996 front-page newspaper story about Mall of America's newly implemented parental escort policy for teen-agers. They begin with a multiple-element lead and gracefully follow its direction:

On the first night of its escort policy for youths under 16, the Mall of America on Friday became a circus of showmanship, guile, profanity, good humor, resourcefulness, at least two arrests and media searching for a story.[24]

In this next paragraph, the writers add more details:

About 150 security officers and mall workers checked identification cards, escorted some youngsters away from the mall's doors, reunited some children with their parents and patrolled corridors like cops on a beat.[25]

Here, a transitional paragraph helps to introduce and emphasize a good quote:

Maureen Hooley Bausch, associate general manager, said she was generally pleased with the results.

"Six weeks ago (on a Friday night), we would have had all sorts of security incidents," she said. "Tonight we're just getting resistance to the policy."[26]

The writers then add details about the arrests:

The altercation that led to the arrests broke out about 7:40 p.m. near Bloomingdale's. The incident resulted in allegations of racism and unconstitutional behavior by those apprehended. A mall spokeswoman said the two men in custody later told officials they intended to get arrested to protest the new policy.[27]

In the following paragraph, the writers signal what the rest of the story is about:

> Much of the evening was a series of vignettes rooted in the big shopping center's effort to prevent large groups of unescorted youngsters from annoying shoppers and disrupting normal business.[28]

From here, the story jumps from page one to an inside page, where the writers follow up on their implied promises of more information. One short section of the story highlights various vignettes, another goes into detail about the altercation; still another discusses the businesslike approach of mall security in enforcing the new policy.

In covering the same story for the broadcast media, KARE television reporter Dennis Stauffer cannot be as thorough as the newspaper reporters, because he has only a couple of minutes of air time. Still, he neatly summarizes the highlights of the story using a combination of the tools available to him: good pictures, natural sound (from encounters at the mall) and tight, crisp writing with a conversational tone. As we saw in Chapter 5 (p. 111), a summary lead read by an anchorperson introduces the reporter's story. Stauffer then opens his video track with the natural sound of a mall security officer asking young men for their IDs, which is followed by these comments:

> The approach was polite to an extreme as IDs were checked and the new policy explained.
>
> (Security officer says to a young man, "Have a good evening. Thanks for your cooperation.")
>
> Still, despite all the publicity, a few came unprepared . . .
>
> (A young shopper says, "I didn't bring ID.")
>
> And left frustrated.
>
> (Other young shoppers criticize the policy, calling it "dumb and weird.")
>
> Thirteen protestors criticized the policy but their impact was minimal.[29]

Notice that Stauffer's writing is brief compared to that of his newspaper colleagues. He lets the video tell much of the story, moving slickly and smoothly out of sound bites into his track and back to the sound bites. In doing so, he masters the art of the transition: "a few came unprepared" (sound bite) and "left frustrated" (another sound bite).

Although the broadcast story seems short and crisp, it is wordy, compared to the radio advertisement that Mall of America prepared to introduce its escort policy. The ad is a 60-second conversation between two boys, accompanied by eerie-sounding background music:

(Horror movie music)
Jess: Drew.
Drew: Jess, what's up?
Jess: They're coming to get me, man.
Drew: The seniors?
Jess: No, worse.
Drew: Who could be worse?
Jess: My parents.
(Music Hit)
Announcer: It's out there. And it's coming to get you. It's the Invasion of the Parents.
Drew: What are they gonna do to you?
Jess: They're —
Drew: Yeah?
Jess: They're —
Drew: WHAT, MAN?
Jess: They're taking me to Mall of America.
(More Music)
Announcer: It's amazing. It's incredible. It's Mall of America's new parental escort policy.
Drew: People will see you, man.
Jess: They're on the stairs.
Drew: You gotta get outta there.
Jess: They're coming down the hall.
Drew: Get under the bed.
Jess: I AM UNDER THE BED.
Announcer: Friday and Saturday nights from 6 until closing, under 16 must be accompanied by parent or guardian.[30]

And just in case anyone missed the message, the writers add a crisp warning tag from the announcer:

Announcer: Invasion of the Parents. Under 16 not permitted without parent or guardian. Call xxx–xxxx to find out more. Mall of America. You can't go alone.
Jess and Drew: AAAAHHHH.[31]

Move the Story Along

According to Roy Peter Clark, a nationally renowned writing coach, a good way to keep a story moving is to place "gold coins" along its path, rewarding the reader, listener or viewer for hanging in there.[32] Dennis Stauffer's news director at KARE-TV in Minneapolis/St. Paul, Tom Lindner, wants his producers and reporters to save something interesting for the middle of their stories. Stauffer does just that in his story about Mall of America's escort policy, carefully dropping a gold coin in the middle of his video track:

> The biggest crisis we found came when a 20-year-old handi-capped woman was told her 15-year-old helper couldn't go in. Anyone under 16 must be with someone over 21.
> Then, a good samaritan stepped forward to help.
> (Natural sound from a man who says, "I'll escort her around the mall.")
> Moments later a mall manager agreed to make an exception for her, one the mall didn't anticipate, but one they say they will continue to make.[33]

Similarly, Star Tribune reporters Wascoe and Apgar scatter a few gold coins in the middle of their story on the mall's new policy:

Here are some scenes from around the megamall on Friday night:

Charles Davis, who said he was a 17-year-old student at Bloomington Kennedy High School, was turned away by officers at the east entrance. But as he headed for the bus station, he asked an arriving older man whether he could enter with him to abide by the policy. The man said yes, and Davis was allowed in. The two men immediately went their separate ways. . . .

Julie Martini, a 21-year-old senior at Carleton College in Northfield, Minn., was asked for identification twice: once when she tried to buy a champagne truffle at Godiva's candy shop and again when she re-entered the mall from one of the department stores. She was miffed that officers didn't think she looked her age.[34]

Advertising writers also try to move an ad along with "gold coin" surprises. In the Polaroid ad shown in Figure 6-3, page 146, for example, a photo of a dog ready "to do his duty" on a neighbor's lawn provides the focal point. It is accompanied by a few lines of gritty dialogue between neighbors that bring the hypothetical exchange between them to life in an unexpected way:

> Look, Lady:
> It's not my dog.
> My dog stays tied up all day.
> Besides, my dog went to obedience school.
> No way it's my dog.
>
> Dear Neighbor: It's your dog.[35]

Use Vivid Details and Language

Striking details, relevant facts and witty dialogue help to keep readers or viewers interested in a news story, while bursts of colorful or evocative writing can delight readers or viewers and make their effort seem worthwhile. Language that is colorful and evocative uses strong words and rhythmic sentences.

Choose Strong Words. Use colorful nouns and active verbs. The idea is to paint a word picture, one that draws the reader into the page or glues the viewer to the screen. Think about the action you are trying to describe. Newspapers do not just come off a high-speed press; the press "spits" them out. Rivers may "tumble" or "meander," and people may "amble" or "scurry."

However, evocative verbs must accurately describe the action. In the following example from a 1996 Los Angeles Times background piece on the turmoil in the Balkans, reporter Tracy Wilkinson starts strong and keeps at it:

BELGRADE — As the streets of this capital fill day after day with whistle-blowing demonstrators demanding an end to his rule, Serbian President Slobodan Milosevic has remained largely silent on the subject.

But his wife is a different story. Mirjana Markovic stormed back from a book tour in India, and almost immediately the regime went on the attack. She accused the generally peaceful demonstrators of causing

Look, lady:

It's not my dog.

My dog stays tied up all day.

Besides, my dog went to obedience school.

No way it's my dog.

Lots of dogs run loose in this neighborhood.

I guarantee you it's not my dog.

Dear Neighbor:

It's your dog.

◆ **Polaroid** *See what develops.*

Figure 6-3

"brutal" havoc in the nation; they suddenly became "fascist malcontents" and enemies of the Serbian people. . . .

After her husband, Markovic is the most powerful person in Serbia —

and one of the most reviled. Seen as evil yet tragicomic by her many detractors, Markovic is the favorite target of demonstrators who spoof her girlish dress and ridicule her pretentious airs.[36]

Wilkinson's prose is alive with passion and precision: "whistle-blowing demonstrators" are "demanding;" Mirjana Markovic "storms" back

2/4 2:45 pm

from her trip and starts "accusing" demonstrators; she is "reviled," and her detractors "spoof and ridicule" her.

Similarly, KARE-TV reporter Rick Kupchella uses strong nouns and verbs in a 1996 story about the dangers of house fires. In the lead-in to his videotape story, Kupchella uses parallel phrases to tell viewers that fire is one of the few things in the world that can be "commonly desired and completely devastating." He goes on to write that a fire is reported in Minnesota every 25 minutes and that fires killed 81 state residents in 1995. The most deadly part of any fire, Kupchella notes, is the smoke. Using the advantage of compelling pictures to help viewers understand the problem, the reporter conveys a similar sense of energy and urgency with his words:

> In a serious fire, like this one documented by KARE-11 News just over a year ago, it is plain to see how smoke takes over a home.
> Watch as that curtain of smoke drops before the window across the room.
> It is only seconds before it is pitch black — in the middle of the afternoon. This makes it not only hard to see but hard to think . . . because the same smoke that blinds you also disorients you. . . .
> *(Dramatic pause in the script)*
> . . . Robbing oxygen from your lungs and your brain, too.
> Remember, the smoke will make it so dark in here . . . you can't even see the fire.[37]

Write Rhythmically. The beat or rhythm of a story may vary from sentence to sentence or paragraph to paragraph. Many writers vary sentence length in order to maintain the reader's interest. For example, in the dispatch from Belgrade quoted on pages 145–46, Tracy Wilkinson follows a 30-word, scene-setting lead describing Serbian President Milosevic's silence with a seven-word shift in gears: "But his wife is a different story."

Similarly, in a 1996 Star Tribune story about Kirby Puckett's retirement from professional baseball after a glaucoma diagnosis, reporter Jim Souhan uses a mix of short and long sentences to get his story off to a compelling start:

> The news conference called to announce Kirby Puckett's retirement from baseball didn't start on time. His teammates wouldn't let it.
>
> While a crowd waited Friday in the Halsey Hall Room at the Metrodome, the most popular athlete in Minnesota history strolled in with his wife. Tonya Puckett was crying; her husband wore sunglasses and his white Twins jersey. A gauze patch covered his right eye.[38]

Advertising writers, especially those writing television ads, often pace their scripts by using parallel words and phrases and varying sentence length. In a 1991 television ad for Andersen Windows and Patio Doors, for example, the writers use these devices to create an effective 30-second piece. Light streams through a window, casting shadows on the wall and floor of a home. Then the viewer hears a woman's voice say:

> Light.[39]

Following this one-word sentence fragment, a longer sentence expands on the opening:

> It is the first element of creation.[40]

Next, the writers move into the body of the ad with grace and rhythm — and a couple of parallel phrases — while the camera glides slowly and smoothly through the home:

> Nothing brings it more elegantly or shapes it more beautifully than Andersen Windows and Patio Doors.[41]

Note the rhythmic patterns in "*brings* it *more elegantly or shapes it more beautifully.*" Finally, the viewer is snapped to attention with another one-word sentence, followed by an end to the little story:

> Come. Live in the light of Andersen windows. Come home to Andersen.[42]

The writers want viewers to remember the name of the product and the message in the ad: Andersen windows let in the light.

Create a Memorable Ending

An ending conveys to the reader or viewer a sense of satisfaction: You have come this far so let us sum it up or give you something to think about or remember. Sometimes a simple summary will suffice. The Andersen ad writers use 12 words in three sentences — shortest, shorter and short.

The Mall of America's public relations release uses three long sentences to sum up the mall's experience with the first weekend of its new escort policy. The summary ends with a reminder about why the policy has been instituted:

> "The first weekend taught us to expect the unexpected, like kids showing up for dinner before a homecoming dance," [Teresa] McFarland said. "Our goal is to make the mall safer, not to punish anyone, so we will continue to work with our guests to make sure the policy is fair."
>
> Mall of America implemented the Parental Escort Policy to increase the level of adult supervision in the mall on the busiest nights and to reduce the potential for serious incidents.[43]

The Star Tribune's version of the same story adds, in the concluding paragraph, information about why adult supervision is needed at the mall:

> The escort policy grew out of a series of episodes involving youths at the mall. Merchants complained that groups of youngsters would stand near store entrances and intimidate or discourage shoppers. There were fights and reports of spitting, chases and abusive language.[44]

Some stories, however, demand more than a simple summary. These stories should not screech to a halt; rather, they should roll to a stop while conjuring up an image, prompting a feeling, inspiring a thought or provoking a point to ponder. An advertisement, for example, can leave readers with something to think about. The half-page ad shown in Figure 6-4 appeared in newspapers during Billy Graham's 1996 crusade. It features a small, black-and-white picture of Graham behind a pulpit and two lines of type underneath: "This is it. Your last chance to hear Billy Graham delivering the best news you've ever heard. Billy Graham & Friends • Metrodome • June 19–23 • 7:00 P.M. • Free Admission." But the story is in the three-line headline above the picture. It goes like this:

> Ten to nothing, devil. [The beginning of the story.]
> Bottom of the ninth. [The middle of the story.]
> You're up. [The point-to-ponder ending.][45]

Ten to nothing, devil. Bottom of the ninth. You're up.

This is it. Your last chance to hear Billy Graham delivering the best news you've ever heard.
Billy Graham & Friends · Metrodome · June 19-23 · 7:00 PM · Free Admission

Tonight, Amy Grant

Figure 6-4

A writer may also inspire the reader or viewer with a final thought by using a wry observation, a play on words, even a bit of philosophy. Ken Speake ends his videotape story on the elderly gladiolus breeder (see p. 134) by playing off his subject's last words from a sound bite:

> *("I believe heaven is here, now," the old man says. "Make your own heaven, that's your life's work.")*

Then Speake finishes his script:

> Carl Fischer says he feels privileged that his life's work has been among the beauty and art of gladiolus breeding.
> And he's content that he's made his heaven.[46]

This is a delicate, reflective ending to a memorable story. Speake knows his task is to reward viewers for their time and patience. That begs for a neatly crafted ending with a crisp feel or a gentle thought.

THE NEXT STEP: WRITING WITH VISUAL AND AUDIO IMAGES

Writing basic stories requires careful and economic word choice: Do not write 30 words when 10 will do. Do not use an adjective when you can find a colorful noun. Learning to be spare and precise with words is essential to advertising copywriters, online writers, broadcast reporters and news producers who work in a visual age of television and computer graphics.

Media writers often try to combine words with pictures or sound, using visual and audio images to help tell their stories. Advertising copywriters make their words mesh with visual images, and broadcast journalists use words with the sights and sounds on videotape. All media writers supplement their craft with an understanding of how words and pictures can work together, whether in an annual report, a magazine layout, a video script or on a Web page. In Chapter 7, then, we will examine these various media formats of writing with images and sound.

NOTES

1. Anne Lamott, *Bird by Bird: Some Instructions on Writing and Life* (New York: Doubleday, 1994), 18–19.
2. Sheldon Clay, oral history interview conducted by David Nimmer, Minneapolis, Minn., 16 Oct. 1997. Transcript available at University of St. Thomas Library.
3. Carmichael Lynch Advertising, "The Book of Harley-Davidson. Chapter 16: The Hopelessly Addicted," 1997.
4. Pat Weiland and Cindy Hillger, "Train Crossings," WCCO 12 p.m. news, 11 May 1989.
5. "Post-It Brand Notes Mark Tenth Birthday," *3M News*, 17 April 1990.
6. Alice McQuillan, "Boy, 14, Tumbles Thru Ice, Is Saved," *The New York Daily News*, 20 Jan. 1997, 19.
7. Ibid.
8. Ibid.
9. "Reno Wants More Time for Clinton Probe," accessed 14 Oct. 1997 at http://cnn.com.
10. Ibid.
11. Ibid.
12. Ibid.
13. Ibid.

14. Ibid.
15. Staff, "Harrelson," KARE 10 p.m. news, 3 June 1996.
16. Ken Speake, "Gladiolus Man," KARE 10 p.m. news, 3 Sept. 1994.
17. Anne O'Connor, "Video Shows Rollins Shoved Photographer; Vikings Head of Security Has Another Run-in with Metcalfe," *The Star Tribune*, 1 Nov. 1995, C1.
18. Ibid.
19. Ibid.
20. David Wildermuth, "Scuffle," KARE 10 p.m. news, 31 Oct. 1995.
21. O'Connor, "Video."
22. Wildermuth, "Scuffle."
23. Anne O'Connor, "Vikings Say Security Head Will Be Disciplined," *The Star Tribune*, 2 Nov. 1995, C5.
24. Dan Wascoe Jr. and Sally Apgar, "Mall of America's New Policy Spawns a Flurry of Activity," *The Star Tribune*, 5 Oct. 1996, A1.
25. Ibid.
26. Ibid.
27. Ibid.
28. Ibid.
29. Dennis Stauffer, "Escort Policy," KARE 10 p.m. news, 4 Oct. 1996.
30. Olson & Co. Advertising, Parental Escort radio ad, Mall of America, 16 Aug. 1996.
31. Ibid.
32. Roy Peter Clark, "If I Were a Carpenter: The Tools of the Writer," *Workbench*, 1994, 2.
33. Stauffer, "Escort Policy."
34. Wascoe and Apgar, "Mall," A19.
35. Polaroid advertisement, *Sports Illustrated*, 9 Dec. 1996, 77.
36. Tracy Wilkinson, "Milosevic's Wife at Center of Storm Battering Serbia," *Los Angeles Times* wire story, reprinted in *San Francisco Chronicle*, 27 Dec. 1996, B1.
37. Rick Kupchella, "Get Out Alive," KARE 10 p.m. news, 11 Feb. 1996.
38. Jim Souhan, "Kirby Says Goodbye," *The Star Tribune*, 13 July 1996, A1.
39. Campbell Mithun Esty Advertising, Andersen Historical Reel, "Experience of Light," 1991.
40. Ibid.
41. Ibid.
42. Ibid.
43. "Mall of America Declares First Weekend of Escort Policy a Success," Mall of America news release, 8 Oct. 1996.
44. Wascoe and Apgar, "Mall," A19.
45. Billy Graham advertisement, *St. Paul Pioneer Press*, 23 June 1996, 13A.
46. Speake, "Gladiolus Man."

7
WRITING WITH VISUAL AND AUDIO IMAGES

*T*elevision journalist Bill Moyers: "No one in politics ever understood
better the power of the picture than Michael Deaver. In the competition between the ear and the eye, your judgment is . . ."

   White House aide Michael Deaver: "The eye wins every time. The
producers have to say, 'O.K., we're going to use it.' And we [the White
House staff] sit back after it's on that night and say, 'Aha, we did it
again.'"

— PBS special, "The Public Mind — Image and Reality in America"[1]

"The eye wins every time." In an interview with reporter Bill Moyers,
Michael Deaver, former adviser to President Ronald Reagan, chortles
over the ease with which the Reagan administration was able to
"plant" stories on the television networks' evening news. Deaver tells
Moyers that producers, not reporters, choose the stories for the network lineup; and these producers, he says, are suckers for a visual
story. After all, tele*vision* news, by definition, depends on visual images. No politician was more successful than President Reagan in
courting a positive image from the media: He was photographed with
high school students singing "God Bless America," with skydivers
waving an American flag and with union members spouting the Pledge
of Allegiance. Reagan managed to look presidential amid the fan-
fare — to remain above the fray and in control — using the visual
media to his advantage. Politicians know that since the advent of television in the 1950s, voters have come to pay more attention to visual
images than in the days of radio and newspapers, when most people
based their voting decision on the political party to which a candidate
belonged.[2]

   This chapter focuses on the importance of visual images and sound
in media writing. Specifically, we examine how media writing that

incorporates one or both of these elements differs from writing that relies on words alone. We explain the key points of writing with images and sound, including moving images with sound such as television news videos, created images such as in radio news and still images such as on Web pages and in magazine advertisements.

COMPOSING VIDEO SCRIPTS: MOVING IMAGES

Learning the basics of composing a video script or story is important for all media writers, no matter their specific journalistic focus. For television journalists, the skill is fundamental to the job. Public relations professionals must know how to prepare video news releases, stories generated by an organization that are videotaped and produced like television news and then sent to stations for possible use. Advertising copywriters constantly work with images, oftentimes for television commercials or videotapes viewed directly by customers. And newspaper and magazine journalists also benefit from an understanding of scriptwriting: It provides an exercise in tight, concise writing as well as an appreciation for the constraints under which broadcast journalists operate. While media professionals other than television reporters use video scriptwriting techniques, our focus in this section is on the television news. Learning to write broadcast stories means learning the basics of good scriptwriting: The copy must be concise, active, specific and conversational. Writers can use these tools in other video formats, such as in writing the copy for a television advertisement, video news release or corporate training video.

Video images capture people's attention. Certain aspects of moving images alert the viewer's mind to look and listen, including scene changes, graphic detail, close-up shots and quick pacing. At the same time, studies have found that video images require less effort to process than a printed text.[3] This explains why many people find it easier or more convenient to watch television than read a book, as well as why people commonly use television to learn about unfamiliar topics. Television also has the capacity to reach people's emotions. A riveting novel or newspaper article can make an involved reader laugh or cry just as powerfully, but television can employ a wider range of sensory details — scenery, background images, familiar faces, music and other sounds — making it easier for people to process information on an emotional level.[4] Like the print media, the television news also aims to provide straightforward information, and for a good reason: A 1996 study reported that most people get their news from television.[5]

A video image, therefore, functions to capture attention, create interest, stir emotions and provide information. Understanding these functions can help a writer find and combine the elements necessary for a compelling video story. These elements include the attributes of storytelling, dramatic pictures, natural sound and sound bites.

Use the Attributes of Good Storytelling

All good stories have a beginning, a middle and an end, as we saw in Chapters 5 and 6. This sequence gives the story a flow, rhythm and predictable pattern. It makes the story easy for the reader to comprehend and for the writer to organize. Print journalists and public relations writers sometimes vary the storytelling sequence, writing instead in the inverted pyramid style, which orders information from most to least important (for a review, see the Writing Tips box in Chapter 6, p. 138). A good television scriptwriter, however, usually structures the story with a clear beginning, middle and end. Consider this example from a 1997 WSOC-TV broadcast:

> A man in Sweden finally figures out why his nose is constantly stuffed-up.
> There was a piece of cloth inside his head.
> *(Now pictures appear on the screen.)*
> He was blowing his nose when a piece of cloth came shooting out. It's 31 inches long.
> The man had surgery on a brain tumor last month. It seems doctors accidentally stitched the man up with the cloth still inside.
> It's out now and he's breathing easy.[6]

A good ending is especially important for radio and television stories; as we saw in Chapter 6, listeners and viewers expect these stories to come to a satisfying end or conclusion. The scriptwriter of the stuffy-nose story ends with one short sentence: "It's out now and he's breathing easy." But in another example, television reporter Dennis Stauffer takes a bit longer to sum up a 1996 story about Mall of America's newly implemented parental escort policy. His ending is also sat-isfying and conclusive, but it includes supporting facts and observa-tions:

> So while the new policy required some improvising on the
> first night, and not everyone was happy with it, from what we
> could tell it created no major problems or disruptions — which
> is exactly what the mall hoped would happen.[7]

Conversational Tone. A conversational tone is an important attribute of good storytelling. It involves writing the way you talk, but without using slang or sloppy grammar. A storyteller avoids jargon and instead uses real words, personal words and colorful words that might be heard in a conversation with a friend. A storyteller does not use phrases like "we regret to inform you," "the event impacted his life" or "the authorities apprehended the perpetrator." The storyteller writes instead, "I hate to tell you," "that affected his life" and "the police caught the robber."

Simple Sentences. Using images gives the writer an advantage: Fewer words are needed to tell the story. The writer does not need to tell viewers what they can see for themselves in the video. Instead, the writer's purpose here is to play off the image and flesh out the basic impression provided by the image.

Writing with images requires not only fewer words but also simpler sentences. The best construction is uncomplicated and straightfor-ward: a subject-verb-object sentence. A dependent clause placed at the beginning of a sentence can confuse the audience, obscuring the pri-mary message of the sentence. Dependent clauses also are more diffi-cult to read aloud, which is why they are avoided by broadcast news writers. The copy must be clean, crisp and easy for the anchor to read. For example, look back at the sentences in the stuffy-nose story on page 157. The sentences are generally clear and simple: A man has a stuffed-up nose; a long cloth came out when he blew his nose; doctors left behind the cloth during a recent surgery.

Simple sentences are usually also short sentences. While a short sentence has no precise word limit, it is typically in the 2- to 13-word range, with a long sentence having, say, around 30 words. Broadcast news writing relies on short sentences that can be read aloud in a single breath. The sentences also tend to feature short words, in short

WRITING TIPS

OBSERVING DIFFERENCES IN WRITING STYLE: PRINT AND BROADCAST MEDIA

Experienced broadcast writers often give beginners this advice: Write like you talk. Writing in a conversational style means producing scripts that are easily read — using simple, informal language as well as a universal style that can be understood by listeners and read with little rehearsal by the broadcaster.

Other style differences in print and broadcast writing often show up in the use of nonletters. For example, numbers and symbols such as *$* and *%* are not part of broadcast style. Instead of writing *$1 million*, which is the Associated Press style for print, a broadcast script should say *one million dollars* because it is easier read aloud.

Here are a few other rules to keep in mind:

1. Spell out all numbers eleven and under. The number 11 in print can look like *double-1* to a broadcaster.
2. Write out *thousand, million, billion.*
3. Avoid abbreviations. Instead of *Johnson & Co.*, write it *Johnson and Company.*
4. Use hyphens instead of periods: *F-B-I* or *C-I-A.* This tells the broadcaster to say each letter.
5. Use title first, then name. Write *Secretary of State Madeleine Albright*, rather than *Madeleine Albright, secretary of state*, to alert the listener that someone important is about to speak. Give the person's name and title first, then his or her quote. This avoids the "What did the president just say?" response.
6. Use ellipses (. . .) to indicate a pause.
7. Add pronunciation guides to difficult names: *Hezbollah (hez-boh-LAH).*

combinations. Because writing with images generally entails telling stories, short sentences are useful tools. They allow the storyteller to take a breath or to ponder the image the words accompany. When read aloud, short sentences allow the reader to develop a sense of rhythm, a sense that he or she is telling a story. Note the ease with which the following script, from a CBS broadcast on the 1989 Exxon oil spill in Alaska, can be read aloud:

> This is spawning season. Nature pushes the salmon to lay their eggs before they die. The fish have returned in spite of the oil. But what will happen to the next generation is one of the long-term uncertainties.
> The short-term devastation is well documented.[8]

Similarly, news correspondent Jay Schadler uses short sentences with a storytelling rhythm in a "Prime Time Live" documentary about his hitchhike across America in 1996. He introduces one of the characters he meets on the road in this way:

> The moon sets. The sun rises. And the Merrimack River is dressed in fog. New Hampshire and Vermont lay ahead — and a fellow named Lee Brown.
> *(Brown says, "Okay. Come on in. You're a reporter?")*[9]

Verbs: Active Voice and Present Tense. Writing with images is about conveying action and immediacy, and the words a broadcast writer chooses can keep a story moving, making it brisk and giving it power. Active verbs help accomplish these goals. Passive verbs do not. Lauren Kessler and Duncan McDonald make this point clear in their book "When Words Collide":

> When passive voice is used, sentences are robbed of power. Strong verbs are weakened by this construction, and awkwardness is caused. . . .
> When writers use passive voice, they rob sentences of their power. This construction weakens strong verbs and causes awkwardness.[10]

Verb tense is also a concern in broadcast writing. In print stories and news releases, writers generally use the past tense to describe actions that have already occurred and the future tense to describe events that will happen in the days or months to come. In broadcast news stories, however, writers tend to rely on present-tense verbs to convey a sense of immediacy. In most cases, broadcast audiences see and hear the news as it occurs or within hours of its occurrence.

Active, present-tense verbs are an element of good storytelling. Just imagine Grandpa with a youngster perched on his knee: "And then," he whispers, peering over the top of his glasses, "the eagle swoops down from the top of that pine tree, dives on the little squirrel — but misses. And the squirrel scurries to the safety of his den." The active voice and the present tense also work together to capture viewers' attention at the beginning of a news show, particularly during the summaries of the news to come, called "teases":

> Life gives him lemons. But this kid can't make lemonade. The health department shuts down his lemonade stand.
>
> It says he's breaking the law.
>
> *(Now the picture on the screen shifts from the lemonade stand to rocks on a freeway.)*
>
> Huge rocks crumble from a mountainside, landing in the middle of the interstate.
>
> Detour details at ten.[11]

The following broadcast writer combines storytelling with active voice and present tense to relay details of a Senate hearing on campaign fundraising. The writer does this in just six sentences. The passage takes only 25 seconds to read aloud.

> Accusations fly during Day One of hearings on campaign fundraising.
>
> Senators are looking into possible illegal contributions.
>
> Republican chair Fred Thompson says China illegally poured money into the last U.S. presidential race.
>
> And now it appears former Democratic fundraiser John Huang might testify.
>
> He refused to take the mic *(microphone)* for months. But now senators agree to consider granting him limited immunity.[12]

Parallelism. Powerful writing also comes from the effective use of parallel words and phrases. Parallelism produces the kind of rhythmic writing that lends itself to being read aloud alongside pictures. Parallel structure often works best in triplets; two words or phrases are too

WRITING TIPS

USING PARALLEL STRUCTURE

Parallel structure is a technique writers use to increase clarity. It serves to align related ideas in a sentence by presenting them in a similar grammatical pattern.[1] Look at the parallel words (italicized) in this example:

> According to the reviews, the film "Jerry Maguire" is *funny, uplifting* and *entertaining.*

The sentence uses single adjectives in a repeating pattern to describe the movie.

In the following sentence, parallel phrases are used:

> Eating a well-balanced diet can *decrease hunger, increase energy* and *prevent illness.*

Each of the three phrases contains a single verb and a single noun. Changing this pattern may not alter the sentence's meaning, but it would disrupt its clarity and cohesion. For example, if *prevent illness* is changed to *keep illness away*, the sentence would read like this:

> Eating a well-balanced diet can decrease hunger, increase energy and keep illness away.

The sentence retains its meaning but loses some of its rhythm because of the need to tack on *away* at the end.

Writers can use *signal words* to set up a comparison or sequence in a parallel structure.[2] Look at how signal words are used in the following sentences:

> *Either* we wash our clothes on Sunday night *or* we wear dirty socks all week.

> She bought the Valentine's candy for two reasons: *First,* she wanted to attract his attention; *second,* she intended to capture his heart.

Now look again at the preceding sentences. Can you pinpoint the other parallel elements?

1. Lauren Kessler and Duncan McDonald, *When Words Collide*, 4th ed. (Belmont, Calif.: Wadsworth, 1996), 133.
2. Ibid., 134.

few, and four, too many. Consider this tease, which aired before a commercial break on an afternoon news show in 1996:

> There's still plenty to come on KARE-11 News at Five:
> A cool car . . .
> Some cute cats . . .
> And one gorgeous garden.[13]

Look, too, at this feature, where the broadcast writer uses triple parallel phrases to describe springtime:

> Ah spring, when the leaves have finally dropped;
> when the lilacs are dripping with purple;
> when the ants are licking the sap off the peony buds.
> Ah yuck, dandelions, millions of dandelions.[14]

Specific Sentences. Writing brief sentences — with as few words as possible and in tight combination — does not mean sacrificing precision or detail. For example, if a robber orders a beverage before he assaults the bartender, you will not add unnecessary words by specifying that he orders a beer before the punch. Precision, specifics, details: They rescue a story from becoming flat and lifeless, and they keep the audience interested.

Television reporter Ken Speake makes his living writing the "softer" feature stories, but the hard facts he includes distinguish his features. In the following example, Speake describes an ice-cutting ritual that occurs every winter on a lake in Voyageurs National Park, near the Canadian border:

> Their equipment may appear old. But they've never found a better way to harvest ice. Yup. That's a 1929 Model-A engine on the ice saw. . . .

(Sound and pictures of starting the saw engine.)
They cut the cakes 22 inches long and 22 inches wide. That's standard in this business.
And when the ice is more than 16 inches deep, they get out the hand saws that'll cut an inch of ice with every stroke.[15]

Speake's script is sprinkled with hard-edged tidbits that leave viewers with the feeling they have learned something from the story (for example, that a 1929 Model-A engine cuts ice cakes that measure 22 inches by 22 inches).

Use Striking Pictures

All scriptwriters, like Speake, tell stories. They use short, simple sentences with active verbs and parallel structure. They also respect the images that accompany their prose, weaving their words in ways that expand on the meaning in the pictures. They know the importance of images and sound in the writing process. Gordon Bartusch, a veteran television news photographer who often volunteered to take new reporters on their first story, understood that his pictures were at least as important as the words that accompanied them. "Without the pictures," he would say, "what you've got is radio."

Television reporter David Wildermuth also understands the complementary relationship between words and pictures. In the following feature story about a retiring security officer at a downtown department store, Wildermuth combines graceful writing with striking pictures, including an opening shot of a sunrise over the skyscrapers:

(Opening shot of skyline and rising sun.)
The sun is the skyline's alarm clock — one more groggy morning in Minneapolis.
(Shot and sound of noisy bus pulling away.)
Downtown the streets can be loud. The coffee can be strong. But nothing wakes 'em up like Jim Fuhrman.
(Quick, short, sound bites of Jim greeting people.)
Technically, he's a loss prevention officer. But the three thousand who use Dayton's daily as a morning shortcut through the skyway know Jim is a one-man customer relations department.[16]

Wildermuth brings in another striking scene to end his video story on the retiring officer. He sets it up with an astute observation, followed by a bit of philosophy:

> If the hardest part for most of us is greeting the morning, for Jim, it's letting it go. But the coffee only lasts so long. And good friends have to go to work.
> It's the start of another day.
> *(The story ends with a shot of Fuhrman picking up a life-size, cardboard picture of himself — signed by hundreds of well-wishers — and lugging it across an empty corridor.)*[17]

As the examples from Wildermuth show, good television news writing involves the seamless stitching together of words, pictures and sound — both natural sound and sound bites.

Use Sound to Bring Words and Pictures to Life

Natural sound is precisely what the term implies: the sounds of nature, the background hum of daily life. Natural sound comes in many forms: the rustle of leaves in a fall breeze, the din of traffic on a hot summer night in the city or the buzz of power saws the morning after a spring storm. It is the accompaniment to all the moving pictures caught on videotape. Thus, it is also an essential tool for every television writer or producer who wants those taped images to fill the ear as well as the eye.

Some television reporters begin their stories with about five seconds of natural sound. At other times, the sound is the focus of the story. Such is the case with John Blackstone's feature on the cleanup of Yosemite National Park after a severe flood in 1997. He begins with shots of the majestic park, accompanied only by the gurgle of running water and silence. Then he writes:

> Yosemite National Park has rarely been seen like this: all the beauty, none of the people.[18]

In the next part of the story, the natural sound captures the noise of workers and machines in the park cleanup. And then Blackstone writes:

But even without the crowds, there is little serenity here. Work crews are everywhere.

Yosemite is still being cleaned up and rebuilt after California's devastating January floods.[19]

In a 1996 video news release for Breathe Right nasal strips, the public relations writer begins the video story with the sound of sneezes — big sneezes from four different people. The viewer quickly learns the story is about nasal congestion from head colds or seasonal allergies. The sneezes sound spontaneous, and they look, well, messy. Picking up where the natural sneezing sound dies down, the writer begins:

> A stuffy nose can be miserable, but a recent Food and Drug Administration announcement could be the biggest news for nasal congestion since the hanky.[20]

This is similar to a suspense lead, with just a dash of hyperbole. But the writer quickly nails down the specifics in the next two paragraphs:

> The FDA recently cleared the way for Breathe Right nasal strips to be sold as a temporary remedy for nasal congestion.
> Already popular with athletes and snorers alike, the strips can also make breathing through a stuffy nose easier.[21]

The pictures that accompany these paragraphs of the script include Breathe Right boxes on a pharmacy shelf and one Breathe Right strip plastered across the bridge of the nose belonging to Jerry Rice, a wide receiver for the San Francisco 49ers.

In the Breathe Right public relations video, the writer blends three sounds: the voice of the script's reader, the natural sound of sneezing and blowing, and sound bites, or direct quotations, from three doctors and one nasal sufferer. The ending to the two-minute track is a play on words, as viewers see a mother tucking her son into bed, presumably for a good night's sleep:

> When it comes to nasal congestion, the answer to what's in your nose could be what's on your nose.[22]

Sound bites in television news generally run about 10 seconds, which is about 20 seconds shorter than a sound bite of 25 years ago.

News producers, strapped for time, want to keep their shows moving at a brisk pace.[23] Short sound bites can pack a vigorous verbal wallop, especially when the words tumble from the mouth of an excited eye-witness. Reporter Brad Woodard begins his videotape story of a 1996 fatal plane crash with a sound bite from a farmer who saw it. The farmer's voice is superimposed over pictures of the crash site:

> ("The plane was just ripped apart," the farmer says in the sound bite. "It's a sight that nobody wants to see. There are parts of the plane a half mile from where the rest of the plane's at — parts of the wings lay out everywhere. It was terrible.")[24]

Woodard then starts his written script, being careful not to repeat what the farmer has just said — and said more dramatically than he could write.

> The plane went down in a remote area near Aitkin.
> Surrounded by rice paddies and muck, what was left of the fuselage came to rest in the underbrush — barely visible, let alone accessible.
> Even by all-terrain vehicle, it took rescue crews hours to reach the crash site. Once there, they discovered the Piper Malibu's occupants, all four of them dead — probably upon impact.[25]

Powerful sound bites come not only from eyewitnesses, but also from anyone with passion or a point of view. Christopher Darden, a prosecutor in the 1995 O.J. Simpson trial, had both. CBS correspondent Bill Whitaker crafts his script around sound bites from Darden's closing argument to the jury:

> Whitaker: Christopher Darden played temperate teacher to Cochran's passionate preacher . . . turning down the heat, raised yesterday to a fevered pitch. And he called on the jury to deliberate with passion, but reason.
> (Darden says, "Take that common sense that God gave you back in the jury room. Don't let these people get you all riled up, all fired up, because Fuhrman is a racist. Racism blinds you.")[26]

Next, Whitaker smoothly writes his way out of the sound bite, acknowledging what Darden has just said:

> Whitaker: He played 52 pick-up, trying to restack the deck of race cards Cochran tossed around court yesterday. The sins of a racist like ex-detective Mark Fuhrman, he suggested, don't wipe out the sins of O.J. Simpson.[27]

Broadcast writers must introduce sound bites with care, making sure the last words of the reporter flow smoothly into the first words of the sound bite. Next, the reporter must make a careful exit, perhaps by picking up a key word or phrase from the sound bite and putting it in the script — but never repeating the essence of what the speaker said.

CREATING IMAGES IN THE LISTENER'S MIND: RADIO SCRIPTS

Radio newscasts, of course, have no visual imagery. Yet radio has long used many of the same tools borrowed by television journalists to create vivid images in listeners' minds. In 1922, Walter Lippmann wrote that our understanding of distant places and other cultures was based on the media's portrayal of them; he said the press could paint "pictures in our heads of the world outside."[28] The mass media of Lippmann's time were primarily newspapers and radio, and while newspapers contained some photographs, the written text provided most of the description. The sound bite, or *actuality* as it is called in radio, puts the voice of the speaker in our homes and on our headphones. Natural sound brings us the buzz of the "eye in the sky" helicopter, providing news on rush-hour traffic. Or natural sound may be the wind and rain in the background of the weather update. A radio reporter who wants to do a story on teen-age culture may go to a popular coffee shop; the natural sound in this case would include the clang of cups and saucers and the hum of conversation.

Radio reporter Dan Olson uses natural sound in a 1997 feature on historic wooden farmhouses. The sound helps listeners imagine themselves on the farm:

(Natural sound of birds chirping in the yard.)

This is Art Wagner's last summer on the farm. The 85-year-old Carver County resident moves to town this year. He leaves behind a museum piece, a relic, a farm home uncluttered with modern contraptions — a house where the old kitchen hand pump still sits by the sink.

(Natural sound of pump handle rasping up and down.)

The square living room is dominated by a lunk of a furnace, idle during the heat of the summer. The oil-burner replaced the wood stove. Wagner says the only heat for the upstairs bedrooms in the winter passes through a round, cast-iron grate around the stove pipe as it rises through the ceiling.[29]

Just as radio and television rely on sound to convey images, they also use similar writing techniques. Communication is predominantly verbal in both formats, and a good writer keeps this in mind, choosing words and sentences that are easily read and understood. The techniques are similar to those described in the preceding section on video scriptwriting: Use as few words as possible, write simple, specific sentences and use active voice, present tense and parallel structure.

WORKING WITH STILL IMAGES: ELECTRONIC AND PRINTED IMAGES

In writing for moving images such as in television, the words are meant to be heard and not seen. In writing for still images, the words are part of the visual element of the story. The audience acquires information not by listening and seeing, but by reading and seeing. In this section on still images, we look at public relations and news writing on the Web, at magazine writing and at advertising writing.

Electronic Images on the Web

When computer users visit a Web site, they are likely to see splashes of color, graphics and icons, headlines and boldface type, and short paragraphs of information — all on the same screen. Dixie Berg, a public relations writer with more than 25 years of experience, says writing for the Web requires more collaboration between writer and artist than other PR formats, such as brochures, newsletters and maga-zines. When developing a Web site for a client, she works with the artist from the beginning of the project, first to develop the site's main concept or theme, and then to discuss the layout, the graphics and the "copy field" — the space available for words. The computer makes writing for the Web different from other types of media writing in two main ways: (1) in the way writing is presented on the screen, and (2) in the way information is organized before writing takes place.

The color, graphics, typography and copy blocks of a Web page are designed to make the page easy for the audience to read. A company sponsoring a Web site wants to keep people interested and to discour-age them from easily clicking to another location. "Reading on a com-puter screen is much more taxing than reading on paper," Berg says. "The key is to break up copy so it gives visual cues, and so it breaks up the gray monotony of the computer screen."[30] A visual cue, for exam-ple, means using a "More" button at the bottom of a screen so the reader can click to the next page. Clicking to the next page keeps the reader active and is less overwhelming than scrolling through several connected screens of text. Writers can "break up" the gray of a page by avoiding columns that are too wide, by using large-sized type, and by double-spacing lines of text. Writing shorter paragraphs and using sidebar stories and "pulled out" quotes also visually divide the screen.

As in most good writing, a well-written Web page depends on thor-ough planning and organizing before the first word is composed. The Web writer begins with the most general information about the orga-nization or topic and then drills down to the specific. "We want to give a quick bite of information and then give people the opportunity to go deeper for more," Berg says.[31] A writer must try to anticipate the questions that readers might ask or the information they want to learn. These topics become the headings on the home page, or navigation page, of the site. Next, the writer orders the information from general to specific. These levels of information become successive pages on the site.

Because a computer screen is a relatively small palette, online writ-ers should avoid wasting words on both navigation and inside pages. The navigation page is the reader's road map, and the writing consists mostly of titles. Words in the titles must precisely describe the content contained in subsequent pages, and the writing on the subsequent pages should consist of simple, straightforward sentences. The writer should reach the point of the story quickly, cutting extraneous words or phrases. Web copy often includes underlined or boldfaced words that serve as hyperlinks to other pages or sites. This instant cross-referencing distinguishes online communication from the other types of media, but, Berg warns, writers should be careful not to overlink. An organization wants the audience to see as much information as possible at the home site, and overlinking may tempt readers to stray too quickly.

Web writing, in general, tends to have a more friendly, informal tone than similar writing in brochures, newspapers or magazines. This may reflect the more interpersonal nature of online communication. However, a writer's tone should always reflect both the audience and the message. For example, information about a new legal research service targeted at lawyers will have a formal, professional tone, even when it is publicized on the Web. This tone will give the message more credibility, and thus more persuasive power, than a message written with slang and jargon.

Web sites, like magazines, may contain several styles of writing — from news and features to opinion and debate. One of Berg's public relations clients is CIBA Vision, a company that manufactures contact lenses and other eye products. Figure 7-1 shows CIBA Vision's navi-

Figure 7-1

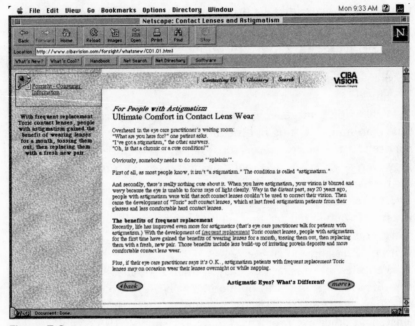

Figure 7-2

gational page, which lists the titles of topics within the section.[32] Note the use of graphics, sidebar information and short text blocks to segment the page. The next page (see Fig. 7-2) is a feature story that is accessed by clicking on the "What's New" category.[33] Like Berg, public relations writers often use a feature style in magazines and brochures. But notice how Berg's story is adapted for the computer screen: Headlines, subheads and several words link to other topics. A sentence from the story is pulled to the side in boldface type, both to highlight the main point of the story and to adjust the width of the copy block. It is also written to fit the screen. Berg makes sure the words are part of the visual element of the page.

These same writing concepts can be found in the online version of Wired magazine, known as Hotwired. The example shown in Figure 7-3 comes from the news section of the Web site.[34] Each item consists of a single, concise paragraph with a short, catchy headline. This helps to break up the screen, as do the graphics and sidebar to the right. The columns do not stretch across the width of the screen; rather, they are narrow, leaving room for white space, another design element that makes online text easier to read.

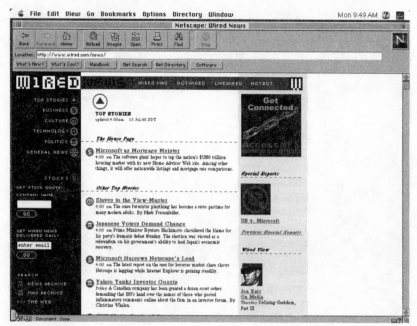

Figure 7-3

Printed Images

Just as copy is part of the overall design of a well-written Web site, words work closely with graphics to convey meaning in posters, billboards and magazine advertisements and articles. Many media writers — including newspaper copy editors and public relations professionals — often combine pictures and words to form a single message.

Magazine Writing. For example, a single message is clear in a People Magazine photo-and-text package (Fig. 7-4) about a woman who wanted to be buried inside her favorite car. The package makes its point about the woman's extraordinary devotion to her automobile without ever saying it directly. How she felt about her 1962 Corvair and what she did because of it are obvious from the three-word headline, a half-page color photograph, a cutline and a smaller black-and-white picture of the woman with her car. The headline is simple: "Long Term Parking." The photograph is memorable: a crane lowering the car, with a casket inside, into the grave.

Long Term Parking

A crane lowered the car with Martin's casket; the widow (below) with her pride and joy in 1999.

Focus When Ralph Nader portrayed the Chevy Corvair as a coffin on wheels in his 1965 book *Unsafe at Any Speed,* he wasn't thinking about Rose Martin. Still, on May 6, Martin, who died at 84 of heart disease, was laid to rest inside her 1962 white Corvair at Tiverton, R.I.'s Pocasset Hill Cemetery.

"She loved her car," says bodyshop owner George Murray, 60, one of two mechanics who had kept the vehicle humming since Martin purchased it new, 36 years ago, for $2,500. "She kept it right up to par." Martin's auto suited her particular needs. A 1970 hip operation had left her with a disfigured right foot, and —as her son Robert, 53, says— the Corvair, which had no hump on the floor, "was the only car she could drive with her left foot."

Over the years, Martin drove the car to her job as a police matron, to the funeral of her husband, Angelo, in 1981 and around town—putting 84,000 miles on the odometer. Robert and his sisters Patricia Boyer, 55, and Agatha St. Amour, 52, first realized the depth of their mother's feelings about the car last year, when she began making arrangements to be buried with it. "We tried our darndest to talk her out of it," says Robert. "But she said, 'No way. That car's going with me.' "

And it did. On a drizzly day, after a Catholic mass, Martin and her car, its original bill of sale still in the glove compartment, were lowered into a grave beside her husband. "She always said she was going to take it with her," mechanic Murray noted. "She wasn't kidding."

COURTESY AGATHA ST. AMOUR

Figure 7-4

Together, the headline and the photo capture the reader's attention. Without belaboring what those elements make obvious, the writer of the accompanying short story begins with a clever reference and adds colorful details to enlighten and explain.

When Ralph Nader portrayed the Chevy Corvair as a coffin on wheels in his 1965 book *Unsafe at Any Speed,* he wasn't thinking about Rose Martin. Still, on May 6, Martin, who died of heart disease, was laid to rest inside her 1962 white Corvair at Tiverton, R. I.'s Pocasset Hill Cemetery.

"She loved her car," says bodyshop owner George Murray, 60, one of two mechanics who had kept the vehicle humming since Martin purchased it new, 36 years ago, for $2,500. "She kept it right up to par." Martin's auto suited her particular needs. A 1970 hip operation had left her with a disfigured right foot, and — as her son Robert, 53, says — the Corvair, which had no hump on the floor, "was the only car she could drive with her left foot."

Over the years, Martin drove the car to her job as a police matron, to the funeral of her husband, Angelo, in 1981, and around town — putting 84,000 miles on the odometer. Robert and his sisters, Patricia Boyer, 55, and Agatha St. Amour, 52, first realized the depth of their mother's feelings about the car last year when she began making arrangements to be buried with it. "We tried our darndest to talk her out of it," says Robert. "But she said, 'No way. That car's going with me.'"[35]

Magazine and newspaper writers and editors increasingly strive for a news or sports package — pictures, headlines and text — that creates an impact. They are trying to compete in a world of images, and the People Magazine package about the woman and her car is a contender.

Advertising Writing. Advertising writers create advertisements that compete with other products, information and activities to capture audience interest; thus, advertising copywriters are under considerable pressure to produce messages that stand out above the rest. The best ads — like the best Web pages — are grounded in a concept. A solid concept also forms the basis of a quality public relations brochure or a top-notch magazine photo essay. While we focus our discussion on advertising, these points can be transported to other media formats that combine still images and words.

Identify the Concept and Audience for an Ad. Creating an effective advertisement does not start with writing a witty headline or shooting a clever photo. The creative process begins first by determining the objectives of the ad and then by devising a communication strategy that will help meet those objectives. For example, in the Nike magazine advertisement shown in Figure 7-5, one objective is to sell sneakers.[36] The communication strategy is to portray the sneakers as tough, and as a product that can help people win. Advertising executives also refer to the communication strategy as the concept of the advertisement, and they say the concept should convey the personality or attitude of the brand being advertised. We could describe the personality of the Nike basketball shoe as tough, aggressive and competitive.

**THE MEEK MAY INHERIT THE EARTH,
BUT THEY WON'T GET THE BALL.**

Just ask Charles Barkley. That's why he likes the Nike Air Force. Especially the cushioning of full-length Nike-Air.® The traction from the Center-of-Pressure™ outsole. And when the league's toughest rebounder thinks the Nike Air Force is the best shoe under the hoop, who's gonna argue?

Figure 7-5

The team that created the Nike ad could have used several other strategies to reach the objective of selling more sneakers. The ad might have simply shown a picture of the shoes and listed their attributes: the quality of the leather, traction and support. Instead, the writers focus on the desires of the audience most likely to buy this product: people who want to play a tough game of basketball, people who want to win, and people who admire basketball player Charles Barkley. Determining the strategy, therefore, requires knowing the needs of the audience and the attributes of the product. This knowledge is based on researching the demographics, lifestyles and needs of the audience, as discussed in Chapters 3 and 4.

Once the audience's needs are targeted and the concept is determined, the creative team thinks of many different ways to execute the concept. The Nike ad features Barkley, portrayed by a specific photograph, accompanied by a certain headline and body copy. The creative director could have chosen another basketball player; the art director another photo; and the copywriter another set of words. How did they decide on the ultimate combination? Amie Valentine, a creative director with 15 years of copywriting experience, says creating a good ad can take 500 attempts at different headlines, pictures and ways to write the body copy. That is, putting together a "perfect solution," as she says, takes a great deal of time: The right words or best headline often does not happen on the 10th try or even the 45th. "It takes 20 tries just to get the clichés out of your system," Valentine says. "By number 88, you're doing better."[37]

Generate Ideas for Copy and Graphics. While writing 500 headlines may seem daunting, advertising copywriters usually do not work alone. They develop ideas with the creative director and the art director, and together they brainstorm for different headlines, graphics and copy approaches. But how does the creative team generate these ideas?

Many experts say that studying good advertising helps; people can try to understand the underlying concept of an ad and how its execution supports that concept. Top-notch ads can be found in Communication Arts magazine's annual advertising edition. The Clio Awards also recognize outstanding advertisements; winning ads can be viewed online at <http://www.clioawards.com>. But Peter Zapf, vice president of the Clio Awards in Chicago, says that copywriters who understand human nature generate the best ideas, and people can learn about human nature by studying arts, literature and history.[38] Notice the headline for the Nike ad: "The meek may inherit the earth, but they won't get the ball." The beginning words are a biblical reference to the Sermon on the Mount (Matthew 5:5). The ending words reflect the personality of basketball player Charles Barkley — the opposite of meek — and connect to the product being advertised. Zapf says the

Nike copywriter had studied philosophy in college — knowledge he may have been able to draw upon when thinking of ideas for the ad.

An advertisement for Cadence Design Systems, a company that helps businesses market products such as pagers and computers, bases its strategy on openness and cooperation.[39] The ad, shown in Figure 7-6, incorporates a knowledge of art into the graphic element: The familiar picture of the all-night diner looks like Edward Hopper's masterpiece "Nighthawks." A closer look at the Cadence ad reveals that the picture has been significantly updated: The man serving coffee has a 1990s haircut and a pager; the woman dressed in red is using a laptop computer; and unlike the original, which shows a man in blue sitting alone at the end of the counter, the Cadence ad's man in blue has a companion.

Copywriters should be familiar not only with historical culture but with current culture as well, says Amie Valentine. She encourages writers to see movies, to read magazines and newspapers, and to develop an awareness of societal trends.[40] One current trend known as *cocooning* or *nesting*, Valentine says, was used with one of her agency's clients, Andersen Windows. The writer incorporated the nesting trend into the ad by showing how the windows make a person feel secure, warm and comfortable. The Andersen advertisement shows a cozy room lit by sunshine through a window (see Fig. 7-7, p. 180). The copy reads: "Television? Are you serious? Rooms like this are made for pondering nature, distant horizons, or even your daydreams."[41] Says Valentine: "If you can create a psychological benefit for the product, and allude to your product's relationship to the trend, you create a competitive advantage for your client."[42]

Copywriters also use other techniques to generate ideas for copy and graphics. One way is to imagine yourself in a room with the ad strategy. The room has six doors, all leading to different ideas that relate to the strategy. Walk through one idea door, and you are in another room with six doors that relate to that idea. Soon you may be able to think of dozens of ideas relating to the strategy. Another way to generate ideas is to think about how you will feel if you use the product or how you will feel if you do not use it. The Wisk detergent "Ring Around the Collar" campaign is an example of a consequence of not using the product that taps into a human need — the need not to feel humiliated at work because your clothes are dirty. Still other ways to generate ideas are to put the product in absurd situations or to change the point of view of the speaker in the ad. An advertisement for

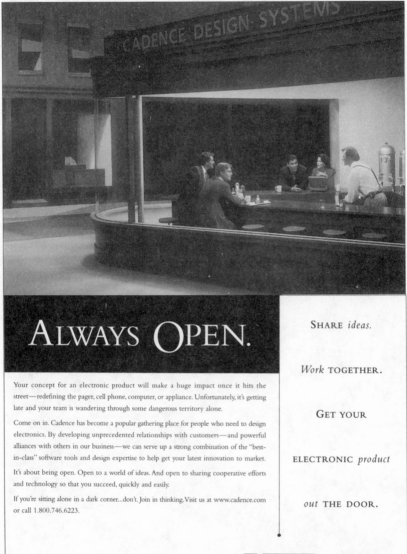

Figure 7-6

Figure 7-7

a household smoke alarm, for example, could be told from the per-spective of a firefighter, a neighbor or even the family springer spaniel. The key point to generating ideas, therefore, is to try as many twists as possible, always circling back to the strategy.

Use a Copy-Dominant or a Visually Dominant Approach. Most good advertisements emphasize either the copy or the visuals, but not both. An ad with an explicit picture works best with a minimum of words, whereas an ad dominated by copy usually needs just a simple picture. The words and the picture work together to complete an idea; the words do not repeat the thoughts in the picture. This is known as the *completion principle:* If either the copy or the picture can stand alone, then one or the other is not doing its job. The ad on page 100, for the National Fluid Milk Processor Promotion Board, is a visually domi-nant ad. It displays a picture of musician Paul Shaffer, who directs the CBS Orchestra on the David Letterman show.[43] The primary words are "MILK . . . Where's *your* mustache?" Instead of simply portraying a glass of milk, the advertisement conveys a personality for the prod-uct. Paul Shaffer is an upbeat, funny and talented musician. He's a "cool" celebrity who seems proud to sport his milk mustache, and the Milk Board links his hip reputation to its product. Other ads in the same campaign link additional celebrities to milk, including movie di-rector Spike Lee, actors Jimmy Smits and Dennis Franz of television's "NYPD Blue," and baseball player Cal Ripkin Jr. of the Baltimore Orioles. Each of the ads includes a few words that reflect the celebrity's personality. Spike Lee urges people to "do the right thing" and drink milk; the "NYPD Blue" detectives ask readers to "reach out" for three glasses a day; and Cal Ripkin Jr., famous for never miss-ing a game, says he never goes a day without milk. Some studies have shown that people transfer feelings about an advertisement to feelings about the product being advertised and that people may act on the basis of these feelings.[44] When readers see pictures of celebrities such as Paul Shaffer, Spike Lee, Jimmy Smits, Dennis Franz and Cal Ripkin Jr. with milk mustaches, the images may evoke positive feelings about the ads; the Milk Board would like readers to feel similarly about drinking milk.

In another advertisement, this one running in The New York Times in 1996, grandparents are urged to talk with their grandchil-dren about drug abuse (see Fig. 7-8).[45] The ad is copy-dominant, and its concept is the powerful relationship between grandparent and

The power of a Grandma.

Children have a very special relationship with Grandma and Grandpa. That's why grandparents can be such powerful allies in helping keep a kid off drugs.

Grandparents are cool. Relaxed. They're not on the firing line every day. Some days a kid hates his folks. He never hates his grandparents. Grandparents ask direct, point-blank, embarrassing questions you're too nervous to ask:

"Who's the girl?"

"How come you're doing poorly in history?"

"Why are your eyes always red?"

"Did you go to the doctor? What did he say?"

The same kid who cons his parents is ashamed to lie to Grandma. Without betraying their trust, a loving, understanding grandparent can discuss the danger of drugs openly with the child she adores. And should.

• The average age of first-time drug use among teens is 13.

Some kids start at 9.

• 1 out of 5 American kids between 9 and 12 is offered illegal drugs. 30% of these kids receive the offer from a friend. And 12% named a family member as their source.

• Illegal drugs are a direct link to increased violence, to AIDS, to birth defects, drug-related crime, and homelessness.

As a grandparent, you hold a special place in the hearts and minds of your grandchildren. Share your knowledge, your love, your faith in them. Use your power as an influencer to steer your grandchildren away from drugs.

If you don't have the words, we do. We'll send you more information on how to talk to your grandkids about drugs. Just ask for your free copy of "A Parent's Guide to Prevention." Call 1-800-624-0100.

Grandma, Grandpa. Talk to your grandkids. You don't realize the power you have to save them.

Partnership for a Drug-Free America.

Figure 7-8

grandchild. Note the simplicity of the photo: a close-up shot of a concerned grandmother and an adoring grandson. The emotion captured in the photo draws the audience to read the brief headline, and, it is hoped, to continue with the body copy. People who are interested in a topic are more likely to read information about it than those who are not interested. Copy-dominant ads should therefore be most effective with people who already have some personal connection to the subject.

Copywriters must also consider the medium in which the ad will appear when deciding whether words or visuals will dominate. The grandparent ad can feature a great deal of copy because it ran in a newspaper, whose audience wants to spend time reading. The audience for a billboard advertisement, in contrast, has less time to read. The ad for Toro snowblowers shown in Figure 7-9 appeared on a billboard.[46] The writers know their audience — motorists — have only a few seconds to read it, and the copy is appropriately short: "Red. White. Blew." The writers put a new twist on the phrase *red, white and blue*, while also incorporating the product's image. Red is the color of the Toro brand; "blew" is the unquestionable result of using the product.

Use Memorable Words. Advertising copywriters should follow a few basic guidelines when choosing their words:

- *Be concise.* Even in an ad dominated by copy, the writer should waste no words. The grandparent ad shown in Figure 7-8 contains many words, but each sentence is tight.
- *Be flexible.* The words and tone of the copy should be adapted to the product being advertised. For example, Peter Zapf's copy for Polaris snowmobiles was written after he had read letters from snowmobile owners to learn the language style of his audience. Not only is this strategy based on consumer research, but the words themselves reflect the speech of the audience.[47]
- *Call the reader's attention to the product, not the ad's execution.* When people see an ad that is too clever, they may tend to remember the ad rather than the product. An advertisement that uses clichés or bad puns may also call attention to the execution and distract the reader from comprehending the main point of the message.

Figure 7-9

- *Give words a lyrical quality.* When read aloud, the copy for an advertisement should flow smoothly like the lyrics of a song or the lines of a poem. Copywriters often read their copy out loud as they edit and refine it, or they ask someone else to read it as they listen and revise.

THE NEXT STEP: WRITING COMPLEX STORIES

Advertising writers often read their copy aloud to listen for rhythmic patterns in words and sentences and to hear the words tell a story. As we saw in Chapter 6, all basic stories have a focus or theme, strong nouns and verbs and grammatically correct prose. Depending on the medium, they also incorporate sound and pictures in the ways discussed in this chapter.

However, a complex story moves beyond the simple carpentry of the basic story and into the realm of craftsmanship. Here, the writer's tools are finer, and the writer's touch is more delicate. A complex story is generally longer than a basic story because it covers a wider range of facts, details, concepts and ideas. It might be a newspaper article that examines the impact of federal welfare cutbacks on single mothers, or a university public relations release that explores the explosion in women's competitive athletics, or a newsmagazine article that describes an assault on Mount Everest in which five mountain climbers froze to death. Because a complex story is a longer journey for the reader, the writer needs to work harder to make the path smoother and the sights brighter. Definitions, descriptions and transitions are important tools in building complex stories. In Chapter 8, then, we will focus on the tools and techniques used in writing complex stories.

NOTES

1. Bill Moyers, "The Public Mind," no. 3 in a series of 4 (New York: Alvin H. Perlmutter Inc. and Public Affairs Television, 22 Nov. 1989).
2. Overall, television has led to a decrease in voters' identification with political parties. In addition to television highlighting a candidate's image, many studies show that people now pay more attention to a candidate's stand on issues

when deciding how to vote. See Steven H. Chaffee and John L. Hochheimer, "The Beginning of Political Communication Research in the United States," in *The Media Revolution in America and in Western Europe*, ed. Everett M. Rogers and Francis Balle (Norwood, N.J.: Ablex, 1985), 267–96.

3. Gavriel Salomon, *The Interaction of Media, Cognition, and Learning* (San Francisco: Jossey-Bass, 1979), 28–87; John Newhagen and Byron Reeves, "This Evening's Bad News: Effects of Compelling Negative Television News Images on Memory" (paper delivered at the annual meeting of the International Communication Association, Chicago, May 1990).

4. Richard E. Petty and John T. Cacioppo, *Attitudes and Persuasion: Classic and Contemporary Approaches* (Dubuque, Iowa: Wm. C. Brown, 1983).

5. Dennis Kinsey and Steven Chaffee, "Communication Behavior and Presidential Approval: The Decline of George Bush," *Political Communication* 13, no. 3 (1996): 281–91.

6. Staff, "Cloth in Head," WSOC 10 p.m. news, 15 July 1997.

7. Dennis Stauffer, "Escort Policy," KARE 10 p.m. news, 4 Oct. 1996.

8. John Blackstone, "Alaska Cleanup," CBS Evening News, 14 Sept. 1989.

9. Jay Schadler, "Looking for America," PrimeTime Live, 6 March 1996.

10. Lauren Kessler and Duncan McDonald, *When Words Collide*, 4th ed. (Belmont, Calif.: Wadsworth, 1996), 77.

11. Staff, "Teases," WSOC 10 p.m. news, 15 July 1997.

12. Staff, "Campaign," WSOC 10 p.m. news, 8 July 1997.

13. Staff, News at Five Tease, KARE 5 p.m. news, 3 June 1996.

14. Ken Speake, "Dandelions," KARE 6 p.m. news, 20 May 1994.

15. Ken Speake, "Ice Harvest," KARE 10 p.m. news, 20 Jan. 1990.

16. David Wildermuth, "Greeter Retires," KARE 10 p.m. news, 25 March 1994.

17. Ibid.

18. John Blackstone, "Yosemite," CBS Evening News, 12 March 1997.

19. Ibid.

20. Shandwick International, "FDA Clearance to Market for Nasal Congestion," video news release, spring 1996.

21. Ibid.

22. Ibid.

23. Jerry Jacobs, *Changing Channels, Issues and Realities in Television News* (Mountain View, Calif.: Mayfield, 1990), 120.

24. Brad Woodard, "Piper Crash," KARE 10 p.m. news, 3 June 1996.

25. Ibid.

26. Bill Whitaker, "O.J.," CBS Evening News, 27 Sept. 1995.

27. Ibid.

28. Walter Lippmann, *Public Opinion* (New York: Macmillan, 1992), 3–20.

29. Dan Olson, "Farmhouse," Minnesota Public Radio, 16 July 1997.

30. Dixie Berg of Shandwick International, interview by Stacey Frank Kanihan, Minneapolis, Minn., 26 June 1997.

31. Ibid.

32. "Forsight: Information on Your Vision," CVWorld (online database), accessed 13 July 1998 at http://www.cvworld.com.

33. "For People with Astigmatism: Ultimate Comfort in Contact Lens Wear," CVWorld (online database), accessed 13 July 1998 at http://www.cvworld .com.

34. "General News," Wired News (online database), accessed 13 July 1998 at http://www.hotwired.com.

35. "Long Term Parking," *People Magazine*, 25 May 1998, 10–11.

36. Nike advertisement, "The Meek May Inherit the Earth, But They Won't Get the Ball," *Communication Arts* 30:7 (1988): 41.

37. Amie Valentine of Campbell Mithun Esty, interview by Stacey Frank Kanihan, Minneapolis, Minn., 19 June 1997.

38. Peter Zapf of the Clio Awards, telephone interview by Stacey Frank Kanihan, 23 June 1997.

39. Cadence Design Systems advertisement, "Always Open," *Business Week*, 30 June 1997, 32H.

40. Faith Popcorn and Lys Marigold, *Clicking: 16 Trends to Future Fit Your Life, Your Work and Your Business* (New York: HarperCollins, 1996).

41. Andersen advertisement, "Television? Are You Serious?" *Traditional Home*, March 1996, 2–3.

42. Valentine, interview by Kanihan.

43. "Milk: Where's *Your* Mustache?" ad, National Fluid Milk Processor Promotion Board.

44. Terence A. Shimp, "Attitude Toward the Ad as a Mediator of Consumer Brand Choice," *Journal of Advertising*, 10 (1981): 9–15; Andrew A. Mitchell and Jerry C. Olson, "Are Product Attribute Beliefs the Only Mediator of Advertising Effects on Brand Attitudes?" *Journal of Marketing Research*, 18 (1981): 318–22.

45. Partnership for a Drug-Free America advertisement, "The Power of a Grandma," *The New York Times*, 3 Nov. 1996.

46. Toro billboard advertisement, "Red. White. Blew," (Minneapolis: Campbell Mithun Esty, Job No. TOSNOG6001, 1997).

47. Zapf, interview by Kanihan.

8
WRITING COMPLEX STORIES

SERVE HOT.

(OR ONE GOURMET'S VERSION OF COOKIN' WITH CATERA.)

I know people who think Eve served the biblical apple merely because she was pinched for time and didn't have to cook it.

I like to cook.

"Where did you learn to cook?" people ask, like there's a cavern hidden deep under the city where we meet; a secret, benevolent cabal of chefs, practicing the ancient rites of both spice mingling and time juggling. "In the kitchen," I always reply. They eye me with suspicion and disbelief.

I like to cook.

Cooking lets me unleash my creative energies. To me, recipes are just vague guidelines. I enjoy using intuition to know whether I should add more garam masala, zest another lemon, grind an extra nutmeg or two. I enjoy it even more when I'm right.

I like to cook.

Once we had fresh 25-count shrimp, begging to be grilled, and no appropriate marinade in the house. I frantically mingled spices in a manner that would have terrified an alchemist. The result is never the same twice and one of my favorites. (I'll share it later.)

Whoever designed my new car must like to cook too. Because they clearly threw out the typical luxury car recipe, scrambled all our preconceptions and came out with a whole new omelet.

The result is Catera, luxury à la fun, the Caddy that zigs.

The performance ingredients are something out of Parnelli Jones' cookbook: four-wheel independent suspension, big four-wheel disc brakes,

ABS, full-range traction control. And a 24-valve V6 that kicks out 200 horses.

Hey, I like to cook.

The inside is something too. Clearly some engineers spent a lot of time slaving over a hot drawing board. Because they've come up with things like seats as comfortable as any this seat has ever sat in. Something about the "hip points" being raised for optimum driving position.

Even the glove box, the oven of cars that turns chocolate to syrup, is different. It has its own air-conditioning vent, so when the a/c is on, it's cool in there.

Well, dinnertime's coming, and I'm out of saffron. Usually I'd just try to make do, but now I feel like a quick zig to the store. You should see me behind the wheel now.

Man, can I cook.

— Cadillac Catera "Serve Hot" ad[1]

Unlike the ads with eye-catching visuals that dominate advertising today, Cadillac Catera's "Serve Hot" magazine ad follows the standard format of most print advertising: headline, body copy and closing, including slogan and logo. The only visuals in the ad, which ran in 1997, are a boxed marinade recipe on the right side of the page, a Cadillac Catera logo at the bottom center, and a tiny photo of the car in the bottom right corner.

Compared to most ad copy, the Catera copy is long and complex. To convey its message, it relies on traditional storytelling, rather than visuals or a balanced mix of visuals and words. The ad's story is built around a theme or main idea, embodied in the narrative of the creative chef. In most complex writing, and in much advertising generally, the theme determines the content and shape of the story or ad. Information, in contrast, often determines the shape and focus of breaking news stories, newsletter briefs, or basic "reason-why" advertising.

Cadillac wants to convey the message that its Catera is not a typical luxury car but a fun luxury car — "the Caddy that zigs." To capture that theme, words fill the page, as readers are treated to a minitale about a hip chef who zigs when he or she is supposed to zag, who follows instincts instead of recipes, who is imaginative rather than predictable and who, therefore, is nicely suited for a Catera.

LEARNING THE FOUR-PART STRUCTURE OF COMPLEX MEDIA WRITING

The Catera ad copy, though relatively short in comparison with the lengthy complex writing of journalists and public relations writers, conforms to the broad, four-part structure common in this type of media writing. The four parts — extended lead, transition, development and ending — are often referred to as *The Wall Street Journal formula*, because the structure is followed by writers of that paper's highly regarded front-page news features (although many other writers, particularly magazine and newspaper feature writers, have also followed the basic structure for decades).[2]

In the Catera ad (see Fig. 8-1) the *extended lead* occupies the first six paragraphs of the copy. The lead introduces us to the chef and narra-

SERVE HOT.
(OR ONE GOURMET'S VERSION OF COOKIN' WITH CATERA.)

I know people who think Eve served the biblical apple merely because she was pinched for time and didn't have to cook it.

I like to cook.

"Where did you learn to cook?" people ask, like there's a cavern hidden deep under the city where we meet; a secret, benevolent cabal of chefs, practicing the ancient rites of both spice mingling and time juggling. "In the kitchen," I always reply. They eye me with suspicion and disbelief.

I like to cook.

Cooking lets me unleash my creative energies. To me, recipes are just vague guidelines. I enjoy using intuition to know whether I should add more garam masala, zest another lemon, grind an extra nutmeg or two. I enjoy it even more when I'm right.

I like to cook.

Once we had fresh 25-count shrimp, begging to be grilled, and no appropriate marinade in the house. I frantically mingled spices in a manner that would have terrified

an alchemist. The result is never the same twice and one of my favorites. (I'll share it later.)

Whoever designed my new car must like to cook too. Because they clearly threw out the typical luxury car recipe, scrambled all our preconceptions and came out with a whole new omelet.

The result is Catera, luxury à la fun, the Caddy that zigs.

The performance ingredients are something out of Parnelli Jones' cookbook: four-wheel independent suspension, big four-wheel disc brakes, ABS, full-range traction control. And a 24-valve V6 that kicks out 200 horses.

Hey, I like to cook.

The inside is something too. Clearly some engineers spent a lot of time slaving over a hot drawing board. Because they've come up with things like seats as comfortable as any this seat has ever sat in. Something about the "hip points" being raised for the optimum driving position.

Even the glove box, the oven of cars that turns chocolate to syrup,

is different. It has its own air-conditioning vent, so when the a/c is on, it's cool in there.

Well, dinnertime's coming, and I'm out of saffron. Usually I'd just try to make do, but now I feel like a quick zig to the store. You should see me behind the wheel now.

Man, can I cook.

MARINADE WITH ZIG

1½ teaspoons garam masala
2 teaspoons brown sugar
2 teaspoons tumeric
1 teaspoon honey
2 dashes soy sauce
½ teaspoon red pepper paste
2 cloves garlic, minced or pressed
½ inch ginger root, peeled and grated
2 tablespoons curry oil

Combine all ingredients. Add shelled, deveined shrimp. Turn to coat, marinate for a couple of hours. Grill with chopped veggies. Eat.

CATERA
THE CADDY THAT ZIGS.

STARTING AT $29,995
TOTAL MSRP OF $30,635 INCLUDES $640 DESTINATION CHARGE.
"Tax, license and optional equipment extra. For the authorized Catera dealer nearest you, call 1-800-333-4CAD or visit us at www.catera.com.

Figure 8-1

tor, who has both a love and a philosophy of cooking. At this point, we are intrigued to know what cooking has to do with a car. The *transition*, which is in the next three paragraphs and which illustrates a unique quality of complex writing, reveals that the same creative instincts that spark the chef's cooking went into creating the Catera, "the Caddy that zigs." The next four *development* paragraphs list various features of the Catera as well as details of interest to new-car buyers. In the final two paragraphs, the *ending* brings us back to the chef, who is now "cooking" in a Catera during "a quick zig to the store."

Most complex stories follow this four-part structure, but the writer does not just fill in the blanks to create the story. Rather, the writer uses the general structure as a guide to organizing a wealth of material in a way that will focus on the theme or main idea of the story. Any one of many variations on the general approach may be developed by the writer. However, in all cases the writer's purpose is the same: to help readers learn about the topic, whether it is an issue, a place, a person, a product or some aspect of the human condition, by taking them beneath the surface of basic facts and information. Let us look now at how each of the four parts of complex writing can work in various ways to achieve that purpose.

PART ONE: THE EXTENDED LEAD

The *extended lead* in a complex piece of writing can use any one of several approaches. We will discuss three: putting a face on an issue, depicting a person, and illuminating the extraordinary by comparing it with the ordinary.

Put a Face on the Issue

The Chicago Tribune covers Mall of America's weekend escort policy for teens in a six-paragraph, 175-word article that provides all the basic information about the 1996 policy. The article combines a thoughtful lead with a classic *who-what-when-where* approach dictated by the information gathered. Here are the first two paragraphs:

A teenage rite of passage — hanging out at the mall on weekend nights — ended Friday when Mall of America began enforcing a curfew for youngsters under 16.

Officials at the nation's biggest shopping and entertainment complex hope to reduce rowdy behavior by requiring young people to be with someone at least 21 on Friday and Saturday evenings.[3]

The article also includes a quote from an activist charging that the new policy is "an attack on youth and minorities."

In contrast to the Tribune's mall account, Robyn Meredith's article for The New York Times takes about 2,000 words to look at some of the issues mentioned but not explored by the Tribune, especially teen rights and racial bias. Rather than letting the facts dictate the form, Meredith builds her article around those themes or issues.

Just as the creators of the Catera ad personalize the selling of the car with a first-person account from a quirky character, the chef, Meredith personalizes the issues raised by Mall of America's new policy by focusing on an individual affected by the policy. In other words, she puts a face on the issue with this lead:

Marcus D. Wilson, 18, has been coming to the Mall of America here once or twice a week since it opened four years ago. He buys tapes, plays video games and sees his friends, especially his girlfriend.

But starting Sept. 28, his habits will be disrupted by the mall's new chaperon policy. People under 16 — including his 15-year-old girlfriend, Stephanie E. Jones — will be barred from the mall Friday and Saturday nights unless they bring a parent or other grown-up over 21.[4]

Focusing on a person whose experience shows readers what the story is about is one of the more common ways of drawing the reader into a complex topic. Such a lead grabs the reader's attention by making an issue human and immediate.[5]

In another example, Washington Post writers Barbara Vobejda and Judith Havemann put a face on the issue of welfare reform by introducing readers to a young single mother:

Tammy Lemieux is a 17-year-old mother with a part-time job and no welfare check. On paper, that makes her a statistical success story, part of the Massachusetts miracle that has driven the commonwealth's welfare rolls to their lowest point in 23 years.

In reality, however, no one — not even the architects of Massachusetts's welfare plan — construes Lemieux's beggarly lifestyle a victory. Kept off the welfare rolls because she refused to live with her parents, Lemieux and her 2-year-old daughter have shuffled from one place to another, staying in a homeless shelter and a hotel, with her daughter's father, a new boyfriend and a sister, in a foster home and a "teen living center" where she got in a fist fight that left her with a fractured eye socket.

Now Lemieux is sharing a dimly lit apartment with the sister of a friend and scraping by on $100 a week from her job at the drive-in window at Wendy's.

"Some teens don't have nobody to turn to," she tells a reporter.[6]

Rather than looking at a big, seemingly incomprehensible issue, the Post writers make the issue accessible and less forbidding by showing readers one specific person instead of many faceless statistics. However, the facts and figures of welfare reform quickly follow in the article, just as ideas about race and civil liberties do in the New York Times article on Mall of America. The Times writer entices the reader with a look at the impact of the parental escort policy on Marcus Wilson and his girlfriend; the Post writers do the same with their tale of Tammy Lemieux.

Similarly, an anecdote can personalize and draw readers into a larger topic. In the following lead from a 1996 newspaper article about a Billy Graham crusade, Leslie Brooks Suzukamo uses this approach:

The hush.

Roger Olson remembers it vividly. The Bethel College and Seminary theology professor was a 12-year-old Nebraskan then, waiting with his parents to see Billy Graham in Omaha at Aksarben Stadium — a dog track whose name was Nebraska spelled backwards.

The stands were crowded and noisy. Then Graham arrived, delivered by a bulletproof, chauffeur-driven limousine. When he stepped up to the platform, there was a hush, Olson recalled. "I remember as a 12-year-old not being restless at all during the entire thing. I remember feeling as if I was a part of something historic, part of something spiritual."

He paused as he reflected upon the memories of his first Graham crusade. "It's one of the reasons I'd like to take my 11-year-old daughter," he said of the upcoming Twin Cities appearance by the 77-year-old evangelist.

The setting this time will be somewhat grander — the Hubert H.

Humphrey Metrodome, Minneapolis' high-tech, air-conditioned, Tupperware-like equivalent of the old-fashioned evangelist's canvas tent.

And the crowds certainly will be larger — they're estimated to reach 50,000 to 60,000 a night for five nights.

But the same cohesion of the crowd, the same desire to be part of something larger than life, the same quest for meaning and sense in daily life will settle over the Metrodome audience as it did over that dog track in Nebraska three decades ago, say scholars such as Olson and organizers of other mega-events.[7]

Suzukamo hooks the reader by focusing on Roger Olson's vivid memory of having been a part of something bigger than himself when he attended a Graham revival as a child. Olson's experience identifies an important attraction of the Graham crusade: meaningful participation in something grand, historical and spiritual. Moreover, Olson's memory of "the hush" ties the crusade to something personal and emotional, and appeals to the reader on that level.

Depict the Person

Although a young single mother is used by the Washington Post writers to illustrate the problem of welfare reform, the story is not a profile of the young woman. If it had been a personality profile, the writers might have used a similar extended lead, but it would have more specifically focused on the person, her apartment, her child, her life.

A character profile commonly begins with a lead that depicts the person in a scene or setting. The subject is shown in action to give readers a glimpse of the person's character, background or way of life. This is how David Finkel opens his award-winning profile of John Lerro, whose life had changed dramatically after a freight ship he was piloting collided with a bridge, killing 35 people. Rather than telling readers how Lerro's life has changed, however, Finkel shows them how through this revealing scene:

NEW YORK — On a gray, miserable April day, a man with a beard is limping toward a boat on the East River. He looks cold. He looks lousy. He has tears in his eyes from the wind and a cast on his left hand from punching a wall. Because of the cast, he can't button his jacket, and it's flapping around in the icy wind.

His name is John Lerro. He doesn't want to be here. He would rather be in bed asleep. Or in sun-blessed Florida. Or in the arms of a beautiful woman. Or even eating breakfast. But it is Tuesday. On Tuesdays he has to be on the water by 8 a.m.

So he has awakened before he wanted to, thrown some bran in a bowl, added some raisins and wheat germ he keeps under his bed, watered the mixture down in the bathroom sink, and stumbled into the unwelcoming arms of a new day.

"I feel terrible," he says. He is limping because of multiple sclerosis, and the wind gusts are playing havoc with his balance. He says, "I look like a drunk."

At the dock, six students from the State University of New York's Maritime College are waiting for him. All of them are glad when he arrives. All of them are anxious. Today's the day they're going to learn how to dock a boat.

And Lerro — John Lerro of the Sunshine Skyway Bridge disaster, *the* John Lerro — is going to teach them.

Taking his seat on the boat, he has only one piece of advice to offer. He is the voice of experience, and the students are attentive.

He says, "If you misjudge, you've got hell to pay."[8]

In this lead, Finkel reveals the nature of Lerro's life and lays the groundwork for expanding his picture of the subject by exploring why Lerro says just after the lead that he's "living like an animal" five years after the Skyway Bridge collapse.

Newspaper reporter Rosalind Bentley's 1996 profile of an African American woman provides another example of an effective extended lead. The writer opens with her subject, a prominent civil rights activist in three states, in interaction with others. But notice how Bentley incorporates African American expressions and slang into her description of the encounter. Using slang and dialect effectively and respectfully is not an easy task; the writer must have a good ear, a subtle touch and an understanding of the audience. Here is Bentley's lead:

As older black people steeped in the struggle might say, it was clearly one of those "my people, my people" moments. The kind that makes a good, dignified woman or man shake her or his head in puzzlement over why their people have to cut the fool in front of white folks.

Coasting down the escalator at Dayton's one autumn afternoon in all their teenaged glory were four black girls not long out of seventh period class, just a-cussin' and a fussin'. *Real loud.*

Two steps behind them was Josie Johnson. Smartly dressed, not a hair out of place. Something about their exchange didn't sit right with her 65-year-old ears or her good, dignified, steeped-in-the-struggle, race-woman sensibilities.

Certainly she hadn't spent the better part of her life trudging up the civil-rights mountain so that little black boys and little black girls could run through the mall showing out. Didn't they understand that black folk still can't afford to talk loud in public, no matter how benign the conversation? Stereotypes have a long shelf life,

and in the game of racial politics image is still everything.

And so here were these four otherwise lovely, intelligent sisters, just rolling around in it. In front of white folks. *Real* loud.

"Girls! Now you all are just *too* beautiful to be talking like that," admonished Johnson, her arms outstretched in the gimme-strength-Lord position.

The teens' eyes widened and their hands reflexively cupped over their mouths.

"Ooooo, we're sorry," the quartet sang small.

Right then Johnson knew there was still hope for the race.[9]

Bentley dramatizes a moment on a department store escalator to show readers a dignified yet passionate woman who is of and for her race. Like other good storytellers, Bentley weaves facts into the narrative and establishes a strong voice and point of view through a mix of rhythm and pacing, repeated words and phrases, and dialect.

Illuminate the Extraordinary with the Ordinary

Enticements come in many forms. For instance, International Paper highlights its role in providing the country with fruits and vegetables out of season with an ad that juxtaposes the familiar with the less familiar, the ordinary with the extraordinary, to capture the reader's eye (see Fig. 8-2). The ad's headline reads, "Cherries in winter, peaches in spring. It's not the weather that's changed, it's the packaging." The first sentence of the ad then says, "Every day, at 36,000 feet, a global exchange of sorts takes place."[10] Together, the headline and the first sentences serve the same function as an extended lead: They introduce the theme or main idea, in this case, that it is now possible to provide out-of-season fruits, not because the growing season has changed but because the methods of packaging and shipping fruits have changed. An element of mystery is added with the

Cherries in winter, peaches in spring. It's not the weather that's changed, it's the packaging.

Every day, at 36,000 feet, a global exchange of sorts takes place.

Millions of freshly picked items crisscross the globe, many of them gently nestled in packaging's version of a first-class seat— a carton or container designed by International Paper.

Chilean grapes land in Marseilles. California melons touch down in Warsaw. Tuscan tomatoes arrive in Kyoto.

What helps them survive the trip? Package design that anticipates reality: temperature swings, humidity, jostling, customs delays, curious spiders and the occasional 15-foot plunge from a cargo ship's hoist.

Packaging also has to be specific. Frozen chicken, fresh juice, fine china— each poses a very different challenge.

Last year alone, our engineers designed over 44,000 distinct kinds of packaging for businesses all over the world. And in a lab where we mimic the rigors of global travel, our packages are tested until they reveal their every strength and weakness.

We do it for our customers, and for all of you who crave fresh, unbruised cherries in midwinter.

INTERNATIONAL (A) PAPER
We answer to the world.
www.ipaper.com

Figure 8-2

reference to a "global exchange" at 36,000 feet, contrasting with the more familiar image of cherries and peaches. The seeming incongruity of "cherries in winter" is visually presented in the ad's photo of a man bundled against the cold, snow dusting his heavy clothing. A bowl of cherries, obviously fresh since they still have stems, sits on one hand, while in the other hand he proudly holds a cherry by its stem. The

photo is designed to intrigue readers with its uncommon image, to entice them to read the ad's message, but in the process it reinforces the ad copy itself.

An especially striking sense of mystery and contrast is found in a 1996 Newsweek article. In this cover story about climbers who die or disappear on their way to the top of Mount Everest, magazine writers

Jerry Adler and Rod Nordland create a sense of place by focusing on the mountain's summit (the goal of the climbers) and by using familiar comparisons. Here is their lead:

> It is an area about half the size of a living room, a broken platform of rock and ice nearly six miles up in the sky. Higher than most airliners fly; so high that it sits, most of the year, in the jet stream itself and storms blow in at 100 miles an hour. It takes more than two months to walk up to it, but once there nobody stays for more than an hour or two, because if you reached there in the first place you probably used up most of your luck with the weather.[11]

Although in most cases using *it* in a lead can be awkward and imprecise, its use here works well because the story's focus is obvious from the accompanying headline, subheading and large color photo and because the word contributes to the mystery and awe of the inanimate chunk of rock and ice known as Mount Everest. The basic description and familiar comparisons ("the size of a living room," "higher than most airliners fly") allow readers to imagine a place they have never been and probably will never see, a place that is both simple and stunning. Comparison and contrast illuminate the extraordinary: No more than a platform of ice and rock about the size of a living room, the summit is six miles up, takes two months to reach by foot, and is in the jet stream, where 100-mile-per-hour winds are common.

In another example, this one from a 1981 Wall Street Journal article on the Pacific Northwest logging industry, William Blundell's lead deals with the extraordinary, but here the writer speaks to an audience of businesspeople in terms they will understand:

> Let us say that you work in an office building with 1,000 people and that every day at least two are hurt on the job. Some suffer such ghastly wounds — multiple compound fractures, deep cuts severing muscle, sinew and nerve, shattered pelvises — that they may never return to their old posts. And every six months or so, a body is taken to the morgue.
>
> Almost anywhere, this would be called carnage, and a hue and cry would be raised. But in the big-tree logging woods of the Pacific Northwest, it is simply endured with what logger-writer Stan Hager has called "proud fatalism," and few outside the loggers' trade even know of it. Miners trapped behind a cave-in draw national media attention, but in the dim rain forests men fall singly and suddenly. There aren't any TV cameras.[12]

Blundell uses his imagination to construct a striking and grim analogy that will shock his readers and entice them to keep reading about a group of workers with whom they have little or nothing in common.

PART TWO: THE TRANSITION

As we have seen, the extended lead can involve any number of approaches, depending on how the writer chooses to entice readers', listeners' or viewers' interest in the topic and story. However, the lead must also help ease readers into the body of the piece, guiding them to the *transition*. In this paragraph or section, the writer reveals what the story is *really* about, establishes a plot and theme, and makes the story's point clear. Often called the *nut paragraph* or *capsule statement*, the transition serves as a bridge between the attention-getting extended lead and the body of the complex story.

Move from Lead to Body

A transition can move readers from the story's extended lead to its body in just one paragraph or it may fill several paragraphs. For instance, following the lead we saw earlier to a New York Times article on Mall of America's escort policy, Robyn Meredith clearly lays out what her story is about in a single transitional paragraph:

> The new policy here touches on many serious social issues: safety, race relations, parental responsibility and civil liberties. Malls have always been magnets for teenagers, but rising levels of juvenile violence have put pressure on shopping centers to limit who walks in the door.[13]

In contrast, the lead to Suzukamo's article on the Graham crusade is followed by a transition that is two paragraphs in length. Recall from the lead (on pp. 194–95) that Suzukamo opens with an anecdote about a person's childhood experience attending a Graham event. Here, in the transition, he tells readers that the same forces that had brought crowds to hear Graham 30 years ago will attract even bigger numbers now to the Metrodome, to what will also probably be Graham's last appearance in Minnesota. But Graham's crusade is, like a

WRITING TIPS

WRITING EFFECTIVE TRANSITIONS

The reader's journey through a complex story should be as smooth as possible. Transitions between ideas, between sentences and between paragraphs should flow easily in order for the overall piece of writing to have clarity and coherence. Smooth transitions are especially important in guiding readers through a complex piece of writing.

You can tie ideas, sentences and paragraphs together in your writing by using the following strategies:

1. Repeat key words, phrases or names that connect readers to a preceding paragraph or sentence.
2. Use transitional words, such as *but, and, nevertheless, also, however, consequently* and *similarly.*
3. Incorporate transitional phrases, such as *for example* and *as a result.*
4. Use words that refer to time, such as *since, then, next, after, before, now, later* and *earlier.*

Be sure to choose transitional words and phrases carefully and precisely, and to use them in legitimate connections. For instance, use *but* or *however* to alert readers to something that is contrary or contradictory, and use *similarly* or *also* for things that are alike.

In the following sample transitional passage, *the bus* and *Jones* are repeated from a previous paragraph to connect ideas and move the new paragraphs along. In addition, transitional words and phrases (shown in italics) make the passage easy to read:

> Police said *the bus* hit *Jones* when he broke free and tried to escape by running across the busy street. *But* two witnesses claim that police threw Jones into the street.
>
> *One witness,* the owner of Mighty Clean Laundry, said she looked out. . . .
>
> *Earlier in the day,* a bartender at Main Street Chat-n-Chew had refused to serve Jones. . . .

sports event, a cultural mega gathering, Suzukamo says. Here are his two transitional paragraphs:

> And each night the audience will wait for The Word to fall, probably for the last time in this area, from the lips of a man some judge to be the most successful Christian evangelist since Paul. The desire to be part of history will be strong, observers say.

> It is these forces and impulses that draw people out of their homes and into stadiums for what can only be termed mega-events, events that overwhelm individuals and whole communities and infuse them with energy and purpose.[14]

Together, the extended lead and transition establish the subject and theme of the article and show the reader what direction the piece will take in the body.

Following the lead to the Washington Post piece on Massachusetts welfare reform (see pp. 193–94), the writer uses a three-paragraph transition:

> Across the country, welfare caseloads are plunging: down 34 percent in Massachusetts over the past four years, 41 percent in Wisconsin, 37 percent in Oregon, 36 percent in Oklahoma.

> President Clinton boasts that nationwide caseloads are down 2.1 million people, the largest decline in history. Those numbers, say Clinton and dozens of governors, are the most glaring proof that welfare recipients are moving into jobs.

> Those claims of success have set the tone as control of welfare shifts from Washington to state capitals. But reducing the rolls and saving money may well prove to be a flawed measure of success that gives little indication of whether millions of adults now on welfare are able to climb into the ranks of the working class.[15]

Establish the Plot and Theme

In the 1996 Newsweek article on climbing Mount Everest, the writers provide a transition in one long paragraph that builds on the lead paragraph we saw earlier (on p. 200). It sets a scene, uses striking imagery to show what can happen on Everest, gives facts and hints at the attraction of such a dangerous climb:

> On a sunny afternoon just over a week ago, climbers at the Everest base camp at 17,700 feet saw the sky over the summit turn an ominous deep

purple, while the handful on top felt the wind pick up with the suddenness of an opened window. Clouds boiled up from the slopes below, where the nearest shelter, a cluster of wind-whipped tents, was a 10-hour walk away in a little saddle called the South Col. Over the next 36 hours, five people would die between the summit and the South Col, and three others, approaching the peak from a different direction, would disappear in the same storm. Others would survive with hands so frozen they clinked like glasses, dead black flesh peeling from their faces. And all so they could stand on that little patch of rock, where you can almost feel the wind of the planet's rotation in your face, the place on Earth that's closest to the stars.[16]

Here, the summit is described as tantalizing and terrifying, a place where the wind and cold kills, a place of death and disappearance. Yet the last sentence adds a mystical element to that slab of ice and rock as well as to the climbers' quest: Mount Everest is a place where climbers know the earth moves and where they can reach for the stars. In two paragraphs — the lead and transition — the writers present readers with a plot line (the story of mountain climbers who die on Mount Everest) and a unifying theme (the mystery and madness of wishing to confront death or injury in a place where the earth meets the sky).

We also see a compelling confluence of plot and theme in a 1996 television story by Dave Wildermuth. Its focus is an ongoing investigation into the murder of two young women who had been hiking on the Appalachian Trail in Shenandoah National Park, particularly Julie Williams, whose parents were trying to keep alive the investigation and the memory of their daughter. Wildermuth moves beyond the facts of a basic story to create a human interest piece about the possible meaning of the deaths. After an anchor lead-in that characterizes Julie Williams as a "giving kind of person," Wildermuth begins his tale as the camera provides a panoramic shot of the Shenandoah Mountains:

She lived for the endless beauty outside. Her friends and family found the same inside her.

(Mother talks about her smile.)

Her life, like her smile, was a quick bright flash. To hear the story, she would help anyone, go anywhere.

If something was beyond Julie Williams' reach, no one

would bet it was her birthday. She was 24 years old. Her father, Tom, couldn't know she would never see 25.

(Father talks about Julie laughing about growing old.)

She was murdered on a camping trip along the Appalachian
Trail.

3,000 feet up in Virginia's Blue Ridge Mountains, what's
right with America and what's wrong with it collided.

She was beginning a life with 28-year-old Lollie Williams.

They died together, throats slashed, wrists bound.

When search teams found them at their back-country camp-
site in Shenandoah National Park, they'd been dead for days.

People come to parks for peace and tranquility. And certainly a crime like that shatters that covenant.[17]

For his piece to resonate, Wildermuth knows he has to establish a
connection between viewers and Julie Williams; for his viewers to appreciate her death, he knows they have to appreciate her life. So his
lead starts with her and her parents' poignant reflections. But then he
moves to the murders of Julie Williams and her companion as sym-
bolic of the tension between peace and violence in America. The sad
irony of these good, young people being brutally murdered in the
tranquil wilderness of a national park resonates with viewers as well.

Make the Point Clear

Writing a complex article for a specialized publication, such as an
annual report or a company magazine, as public relations writers often
do, is similar in many ways to writing a complex piece for any other
publication. It entails gathering and sorting material, finding a focus,
and shaping the story into a clear, readable form, one with a lead and
transition that grab the reader's attention and that make the point of
the story clear. The main difference here is that the focus of public relations writing is determined by the purpose, ideals and needs of the
client or organization for which it is written.

Good examples of such complex writing can be found in the publications of special interest groups, such as religious, business or environmental organizations. In *Sierra*, a magazine published by an environmental organization called the Sierra Club, a 1997 article on the
effects of logging makes its point clear: Companies engaged in the

business of clearcutting trees on slopes endanger not only wildlife but human life as well. The writer, Patrick Mazza, pulls no punches in his lead and transition:

> Last November, a drenching downpour on a 60-degree slope outside Roseburg, Oregon, unleashed a wall of mud, water, and rocks that flattened Rick and Sharon Moon's house, devouring the couple and two friends. In the same storm, a mudslide from another clearcut shoved a car off Highway 38 into the Umpqua River, killing Delsa Hammer of Coos Bay.
>
> For the second winter in a row, the Pacific Northwest was pounded by exceptionally heavy rainfall. From Washington to Northern California, rivers bulged over their banks and rain sluiced off saturated land, bringing down whole hillsides and killing a dozen people. In five of those deaths, clearcutting was plainly to blame.
>
> In a part of the country that has long been dominated by the timber industry, the deaths managed to bring home the devastation of clearcutting in a way that collapsing populations of salmon and owls have not.[18]

As in other types of complex media writing, this article begins with a general focus — storm devastation — in a lead that incorporates brief, attention-grabbing anecdotes. But the writer then makes it clear that the article is not about storm devastation or its possible causes. The purpose here is to attack clearcut logging, which Mazza clearly casts as the villain in this tale: "In five of those deaths, clearcutting was plainly to blame." From there, Mazza moves on to the third stage of complex writing, development, where he elaborates on his premise.

PART THREE: DEVELOPMENT

The bulk of a complex piece of writing consists of the *development* or elaboration of the theme introduced in the extended lead and transitional paragraphs. The writer may choose one or more techniques for developing a theme, such as incorporating relevant or supporting facts, anecdotes, quotations and visuals, reconstructing a scene, using a character profile or highlighting important data in lists.

Use Relevant Facts, Anecdotes, Quotations and Visuals

Let us consider how some of the writers we have already looked at in this chapter develop their main idea by citing relevant facts, anecdotes, quotations and visuals. In his newspaper article on the Graham crusade, for instance, Leslie Brooks Suzukamo includes quotes from people he interviewed and facts from the material he read on the topic. Suzukamo then uses this material to develop his theme: the psychological power of such huge rallies. He also points out that today's mega-events are primarily limited to sports and religion, are guided by charismatic leaders typical of mass movements (Graham is linked to Gandhi and Hitler) and may have an uneasy relationship with mainstream society, as the evangelical Graham crusade does with mainline denominations.

Similarly, in the "Mountain Murders" human interest story, Wildermuth carefully takes his TV viewers through the facts of the investigation while also providing more personal information about the victims and their hike along the Appalachian Trail. In addition, Wildermuth includes photographs that the young women had taken during their camping trip, obtained from a camera found at the murder scene. The photos are visual facts. They depict the campers enjoying themselves amid the spectacular beauty of Shenandoah National Park, emphasizing the incongruity of the brutal act of murder.

As the preceding examples illustrate, the body of a complex piece of writing gives the audience insight into the subject. Facts, details, anecdotes, quotations and visuals that explore the theme or move the plot along are strung together coherently. The careful selection of these elements is crucial to the success of a complex piece, and any item that does not contribute to the development of the theme should be omitted during the selection process. Note how Suzukamo's piece on the Graham crusade moves coherently from idea to idea as presented by various experts. Wildermuth's story is pushed along by the information and quotes from investigators, the personal information about the young women, and the photos that follow the path of their hike.

Reconstruct the Scene

As noted in Chapter 6, media writers can keep a basic story moving by placing "gold coins" along its path, rewards for the reader, listener or viewer for staying with the piece. These coins are also essential in complex media writing, where the gold coins may be a series of scenes or mininarratives that contain facts and information about the subject. For example, Wildermuth drops several gold coins into his story when he allows the camera to show viewers the murder site and when he shows and tells about the snapshots of the young women's hike. Much of what makes complex media stories come alive is in the reconstruction of scenes. The details come from interviews conducted by the writer or from the writer's firsthand observations.

Let us look again at David Finkel's profile of the freight ship pilot John Lerro (discussed earlier, on pp. 195–96). About two-thirds through the profile, Finkel reconstructs a scene from his subject's life that took place about a year after the Skyway Bridge accident, using Lerro's own words:

> He remembers his first trip. "The captain was wearing a Mickey Mouse shirt. I'll never forget that. By the time we got to the bridge, he knew something was up. He said, 'You're the guy who hit the bridge.' He said, 'Don't worry. I'm a born-again Christian. This was meant to be.' He was very relaxed. He sat there with his Mickey Mouse shirt, and I kept driving."[19]

Create a Character Profile

Scenes and minitales are particularly important elements in the in-depth profile. A character profile contains background facts about the subject, perhaps even a brief history of the person told through personal recollections, and quotations and anecdotes from other people in the person's life. It should also be filled with scenes similar to the one Finkel uses in his profile of John Lerro, which allow readers to see the person in action and to hear the person speak, not just to the writer but also to friends, family, co-workers and strangers.

By the time readers reach the end of a profile, they should have a good idea of the person's character and life. Look back, for instance, at the lead to Rosalind Bentley's profile of African American activist Josie Johnson, on pages 196–97. In the body of that newspaper article, Bentley develops her profile of Johnson as a bright, hard-working, dedicated person who cares about her appearance and who has made a difference in the lives of others. Yet Bentley never says "Johnson is bright" or "Johnson is hard-working." Rather, she lets Johnson's character and accomplishments, as well as the quotes from people who know her, speak for themselves in creating the portrait.

Use Lists to Highlight Statistics and Other Data

In some complex stories, rather than incorporating a range of facts into the narrative, writers present the facts to readers or viewers all at

WRITING TIPS

USING SIMILES AND METAPHORS

"It topped the hill like a high-speed sunrise."[1] That is how one writer describes a bright orange Corvette racing over a hill with the police in pursuit. The scene might have been described as "an orange blur," but the comparison of the car to something very different (a sunrise) creates a more striking and memorable visual image.

This type of comparison is called a *simile*. It uses a word such as *like*, *as* or *as though* to compare two seemingly different things. John McPhee tells readers that a turtle is "about fourteen inches long and a shining hornbrown"; then, instead of describing it as having "bright spots along the margins of its shell," he uses a simile to say the spots are "like light bulbs around a mirror."[2]

A *metaphor*, like a simile, makes an implicit comparison. But the comparison is more direct than in a simile; instead of saying the orange car is *like* a sunrise, the writer says the car *is* a sunrise. Note how reporter Richard West, for example, uses metaphors to describe the patrons dining in the upstairs and downstairs of a restaurant: "Compared with the babbling brook downstairs, it [the upstairs] is a dawdling river along which elegant and weathered old boats temporarily dock in safe ports."[3]

When used properly with other types of descriptive writing, similes and metaphors clarify and breathe life into the common or familiar, allowing readers to see, hear, smell and feel.

1. James Kindall, "A Shiny Orange Obsession," in *Best Newspaper Writing, 1984*, ed. Roy Peter Clark (St. Petersburg, Fla.: Poynter Institute for Media Studies, 1984), 87.
2. John McPhee, "Travels in Georgia," in *The Literary Journalists*, ed. Norman Sims (New York: Ballantine, 1984), 30.
3. Richard West, "The Power of 21," in *The Literary Journalists*, ed. Norman Sims (New York: Ballantine, 1984), 222.

once, in a listing or summary. Print writers can set off statistics or other data with typographical elements such as a list with bullets, while broadcast writers can present a similar list in an on-screen graphic.

Recall, for instance, our earlier discussions of a Washington Post article on welfare reform in Massachusetts (pp. 193–94 and p. 203), in which the writers introduce a number of statistics in the lead and transition. In the body of the piece, however, they use a bulleted list to put their account in perspective:

... It is clear from the Massachusetts experience that the declining caseloads raise new questions when matched with other statistics:

- Only about half of the adults who left welfare have found jobs.
- Only 13 percent of the 2,000 job slots set aside in private industry for former welfare recipients have been filled.

- There has been no increase in the number of people seeking the Medicaid the state guarantees to help former recipients cover their health care costs as they move into the work force.
- There has been no increase over the past year in the number of people using the state's transitional child care assistance.[20]

The writers then return to their subject, a teen-age mother, whose story brings the statistics to life and serves as a unifying element in the article.

Recall also the International Paper ad shown in Figure 8-2. In the body of the ad copy, a transitional sentence tells readers about the company's role in getting fresh food moved around the world: "Millions of freshly picked items crisscross the globe, many of them gently nestled in packaging's version of a first-class seat — a carton or container designed by International Paper." Next, the writers elaborate on that comment by listing some examples:

Chilean grapes land in Marseilles.
California melons touch down in Warsaw.
Tuscan tomatoes arrive in Kyoto.
What helps them survive the trip?
Package design that anticipates reality: temperature swings, humidity, jostling, customs delays, curious spiders and the occasional 15-foot plunge from a cargo ship's hoist.
Packaging also has to be specific.
Frozen chicken, fresh juice, fine china — each poses a very different challenge.
Last year alone, our engineers designed over 44,000 distinct kinds of packaging for businesses all over the world.

And in a lab where we mimic the rigors of global travel, our packages are tested until they reveal their every strength and weakness.... [21]

The ad ends by coming full circle, bringing readers back to where they started: "We do it for our customers, and for all of you who crave fresh, unbruised cherries in midwinter." As we will see in the following section, a strong ending is the final component of a successful piece of complex writing.

PART FOUR: THE ENDING

Although some basic media stories come to an end when a sufficient amount of information has been supplied, others tell a story with a beginning, middle and end (see Chapter 6). Particularly in a complex print story and in any broadcast story, a strong *ending* should tie everything together, bringing readers or viewers to a satisfying conclusion or final thought. Ideally, the ending to the story is as strong as its lead. It might include a striking quote or a poignant or ironic observation. It should drive home the theme, bringing the story full circle, or leave readers with something to ponder or imagine.

Bring the Story Full Circle

Like the International Paper ad, the end of the Cadillac Catera ad (see Fig. 8-1) brings readers back to where the ad began, with the chef. Similarly, in the Graham crusade story (pp. 194–95, 201 and 203), Suzukamo brings the reader full circle by ending with Olson, the theologian whose childhood memory of hearing Graham in Nebraska opens the story. At the end, after discussing evangelists' views of Catholics and Lutherans as tending to emphasize mind over heart and intellect over emotion, Suzukamo again quotes Olson: "To leave out the emotional side of religion almost guarantees people will go looking for it." Here are the last three paragraphs:

Olson recalled a visiting German Lutheran pastor, very traditional and formal, who was highly critical of what he considered Graham's emotionalism.

When Olson went to Germany, Olson attended the pastor's church, "and nobody showed up. And he admitted it," said Olson.	"I wonder if he saw the connection," the theologian mused.[22]

With this ending Suzukamo sends the reader back through the story, while also emphasizing the emotional connection of followers to Graham's crusade.

Similarly, Rosalind Bentley's profile of Josie Johnson (pp. 196–97) brings readers full circle, reminding them of Johnson's roots, character and convictions:

Johnson relays a tale about the last time *she* showed out in a department store. In front of white folk. *Real* loud. It was years ago, back home in Houston on a break from college. A saleslady sauntered up to her and her mother and derisively addressed them as "you gals."	Johnson lit into the woman with a tongue lashing you wouldn't believe, until her mother pulled her aside and told her in an ever so courtly manner to close her mouth and stop being so base. Let the salesgirl wallow down there in her ignorance alone.

The ending is reminiscent of the message Johnson conveys to the four girls on the department store escalator earlier in the story: "I want us to be free to be the best we can be," Johnson said. "That's what we fought for in the struggle."[23]

Ponder the Point

The writer of the Sierra article on the dangers of clearcutting (see pp. 205–6) sums up the message of the story with a quotation, one that leaves readers with a point to ponder:

> "Forested watersheds make their greatest contribution to the economy when they deliver clean water, abundant fish and wildlife habitat, and irreplaceable recreational opportunities," says Bayles [conservation director of the Pacific Rivers Council]. "When they are logged, it is economic benefits for a few and landslides and floods for the rest of us."[24]

A complex story can instead end with a more general point to ponder, as Dave Wildermuth's "Mountain Murders" story does (pp.

204–5). The TV reporter closes with a touching scene in the Shenandoah woods, during which Julie Williams' parents visit the site where their daughter had died. Next, the camera moves to a sweeping shot of the mountains as Wildermuth says these unsettling words:

> Somewhere in the beauty that brought her here is the ugly secret of what took her away. On the mountain where two lives disappeared, what's lost now is the truth.[25]

Wildermuth ends where he started, with comments about Julie Williams and her tragedy, while also leaving viewers thinking about the elusiveness of truth and justice.

Let Readers Imagine: Ambiguity

The Newsweek article on Mount Everest climbers (see pp. 199–200, 203–4) provides a good example of an ambiguous ending. It touches on the mystique of toying with death to reach the top of a mountain and of the summit's proximity to the stars. Here are the final three sentences:

> In contrast to the survivors of other great disasters, nobody involved with a death on Mount Everest has to ask "Why?" But even so, the serenity of these climbers and their families seems uncanny. What is it that they see from that place, so near to the stars?[26]

Ending with a question leaves the conclusion appropriately open and the mystery of the mountain's meaning to the reader's imagination.

USING EXTENDED STORYTELLING IN COMPLEX MEDIA WRITING

Most of the Newsweek Mount Everest article is a story, a narrative told in the traditional sense. The writers stitch together information, anecdotes and accounts from survivors and rescuers to construct a tale of what happened on the mountain when five people died, three disappeared and survivors lost body parts to the cold. After the lead, transition and information about climbers and climbing, the telling of the story begins: "It was late on the evening of Thursday, May 9, that about two dozen climbers. . . ."[27] This is a classic "once upon a time" start, and the story proceeds chronologically from there, to the rescue of the last survivor and the ending atop the mountain near the stars. The article also has a solid plot and compelling theme.

Many writers use, and some publications demand, this type of storytelling through the entire piece. Unlike the traditional journalistic approach that moves into a narrative or that uses a scene-setting description in the lead only, writers of extended stories construct a tightly woven narrative account from beginning to end.

The most complex form of storytelling is the mininovel. A masterful example of this kind of writing is Anne Hull's "Metal to Bone," a three-part series appearing in The St. Petersburg Times in 1993. The series is about the attempted murder of police officer Lisa Bishop in a low-income housing project in July 1992. Hull's story did not run until May 1993, a month after a judge and jury had decided the fate of the young man charged with the crime. Rather than reading as a typical complex article might, with the four parts discussed earlier in this chapter, the segments of Hull's series read like the chapters of a short novel, each of which builds on carefully crafted scenes. The story's main participants — the police officer, the teen-ager charged with the crime, and his father — emerge as fully developed characters, rather than sources from interviews. As Hull has suggested, it is a story of a family and a neighborhood as much as of a crime. The first segment opens with a section set in italics:

It was just the two of them, father and son, living in a tiny apartment where the only luster was a gold picture frame that held the boy's school photo.

Their neighborhood stole the young. The father clutched his son fiercely.

"I don't want you making the same mistakes I did," he said, the voice of a thousand fathers.

On July 4, 1992, at exactly six minutes before midnight, the son stepped from his father's shadow. "I

just wanted to be known," he would later say.

For his cold-blooded debut, he picked a police officer whose back was turned.

The sound she heard from the gun would reverberate for months.

Click.

It was the same sound the key in the lock makes as the father comes home now to the empty apartment, greeted by the boy in the golden frame.

A file at the Hillsborough County Courthouse Annex contains all the information pertinent to the case.   *But there is no hint of all the things that were lost on Independence Day.*[28]

With these opening lines, Hull makes it clear that this is not another cliché crime story but a sad story of failed hopes and dreams. Then the writer introduces the cop who heard the "click":

Officer Lisa Bishop's secret to guarding a sleeping city was pretzels. The crunching kept her awake. She'd pull into a convenience store on Nebraska, say hey to the prostitutes near the pay phone and buy herself a large bag of Rold Gold for the long night ahead. Her shift was from 9 p.m. to 7 a.m.

Four nights a week, Lisa clocked in for duty at the Tampa police station on the frayed outskirts of downtown. In uniform she was petite and muscular, like a beautiful action-figure doll, with piercing green eyes and size 4 steel-toe boots. She kept her hair back in a French braid. Even under a streetlight, her skin seemed carved in pearl.[29]

Following this precise physical description, Hull tells readers more about Lisa Bishop's life as a cop and about her family. The writer takes readers along with Bishop as she starts her shift the night of the shooting. By the end of the first segment, readers are also introduced to Eugene Williams, the prime suspect, and his father, Carl Williams.

In the second segment, Hull takes readers into the Williams' lives and includes Eugene's arrest. Readers learn that Eugene is an unlikely suspect, probably a good kid who used some bad judgment when he had had too much to drink. He had never intended to shoot Bishop. By the end of the second segment, however, some doubt is cast on that assessment. The prosecuting attorney, Shirley Williams, has been looking for evidence that Eugene may not be as good as he seems. The second segment ends with the attorney's discovery:

Eugene swore — to detectives, his lawyer, his family, to Lisa Bishop — that he was not the sort of person who would ever intentionally harm someone.

But Shirley Williams knew another Eugene.

She worked at home over the weekend, preparing for the trial. Inside one of the many file folders spread out on the kitchen table was her trump card.

A Tampa police officer had given Williams a crumpled note. It was written weeks before the Lisa Bishop incident and left on the windshield of a car owned by a woman in College Hill. The woman and Eugene often argued.

Don't think I have forgotten about you punk no bitch! I am going to put that .45 pistol on your ass. I'll come to your house.

The note was signed with the name Eugene.

Williams had sent the note to the Florida Department of Law Enforcement crime lab for handwriting analysis.

Bingo. It was a match.[30]

In the final segment, Hull tells the story of the trial and ends with Eugene's sentencing and processing for prison, his father proudly holding the diploma Eugene earned while in jail:

The lawyers snapped their briefcases shut, the clerk pounded documents with a rubber stamp and the judge looked down to sign the paperwork. Carl hung there. Even after everything, he clutched that diploma as stubbornly as he clutched his belief in Eugene.

Eugene was taken from Carl's side by a bailiff who guided him to the inkpad, where each finger was rolled and printed, one at a time, on a sheet of fresh white paper. Handcuffs were snapped around his wrists.

The sound was an unmistakable click.[31]

An epilogue then briefly relates what happened to each of the story's main characters. Ironically, the writer points out, even in prison Eugene continues to talk about someday becoming a police officer, while the police officer, Lisa Bishop, seldom thinks of Eugene but never leaves her back exposed, even when she eats in a restaurant.

This type of elaborate nonfiction storytelling is highly sophisticated and takes years of practice to master. It also requires good editing. But the characteristics of such storytelling, as we have seen, can be applied to media writing generally, and particularly to complex media writing. Writers should always be on the lookout for a theme revealed by the facts, for a plot that allows for the construction of scenes, for details that define character and create a sense of place and for ironies that say something about life — all of which help readers to see.

NOTES

1. Cadillac Catera ad, "Serve Hot," *Architectural Digest*, Aug. 1997, 49.
2. For a discussion of the Wall Street Journal approach, see William E. Blundell, *The Art and Craft of Feature Writing: Based on the Wall Street Journal Guide* (New York: Plume, 1988), especially chapter 5, "Organization," where the author describes the four stages as "(1) Tease me, you devil (Intrigue me a little. Give me a reason for going on . . .); (2) Tell me what you're up to (What is your story really about?); (3) Oh, yeah? (Prove what you just said . . . Let's see your evidence.); (4) I'll buy it. Help me remember it (An ending that will nail it into my memory)."
3. From Tribune News Services, "Mall of America's Teen Curfew Is Called 'An Attack on Youth,' " *Chicago Tribune*, 5 Oct. 1996, 4.
4. Robyn Meredith, "Big Mall's Curfew Raises Questions of Rights and Bias," *The New York Times*, 4 Sept. 1996, A1.
5. Some studies have shown, however, that using an individual to illustrate a social problem can cause the audience watching or reading the story to believe that the individual is responsible for solving the problem. In contrast, news stories that set the problem in a larger social context lead audiences to believe that government, political leaders and society in general have a role in solving the problem. See Shanto Iyengar, *Is Anyone Responsible: How Television Frames Political Issues* (Chicago: University of Chicago Press, 1991).
6. Barbara Vobejda and Judith Havemann, "Success After Welfare?" *The Washington Post National Weekly Edition*, 13 Jan. 1997, 6.
7. Leslie Brooks Suzukamo, "Special Electricity of Crowd Energizes Graham Crusade," *St. Paul Pioneer Press*, 12 June 1996, A1.
8. David Finkel, "For John Lerro, Skyway Nightmare Never Ends," *St. Petersburg Times*, 5 May 1985. Reprinted in *Best Newspaper Writing, 1986*, ed. Don Fry (St. Petersburg, Fla.: Poynter Institute for Media Studies, 1986), 86.
9. Rosalind Bentley, "Bridge Builder," *Star-Tribune* (Minneapolis-St. Paul), 6 Oct. 1996, E1.
10. International Paper ad, "Cherries in Winter," *The New Yorker*, 29 Sept. 1997, 2–3.
11. Jerry Adler and Rod Nordland, "High Risk," *Newsweek*, 27 May 1996, 52.
12. William Blundell, "The Fatal Fraternity of Northwest Loggers," *The Wall Street Journal*, 8 Dec. 1981, 1. Reprinted in *Best Newspaper Writing, 1982*, ed. Peter Clark (St. Petersburg, Fla.: Modern Media Institute, 1982), 13.
13. Meredith, "Big Mall's Curfew."
14. Suzukamo, "Graham Crusade," 1, 5A.
15. Vobejda and Havemann, "Success."
16. Adler and Nordland, "High Risk."
17. Dave Wildermuth, "Mountain Murders," WCCO 10 p.m. news, 12 Nov. 1996.
18. Patrick Mazza, "The Mud Next Time," *Sierra*, May–June 1997, 22.
19. Finkel, "John Lerro," 92–93.
20. Vobejda and Havemann, "Success."

21. International Paper ad.
22. Suzukamo, "Graham Crusade," 5A.
23. Bentley, "Bridge Builder," E8.
24. Mazza, "Mud," 24.
25. Wildermuth, "Mountain Murders."
26. Adler and Nordland, "High Risk," 57.
27. Ibid., 53.
28. Anne Hull, "Metal to Bone Day 1: Click," *St. Petersburg Times*, 2 May 1993. Reprinted in *Best Newspaper Writing, 1994*, ed. Christopher Scanlan (St. Petersburg, Fla.: Poynter Institute for Media Studies, 1994), 3.
29. Ibid.
30. Anne Hull, "Metal to Bone Day 2: Mean Streets," *St. Petersburg Times*, 3 May 1993. Reprinted in *Best Newspaper Writing, 1994*, ed. Christopher Scanlan (St. Petersburg, Fla.: Poynter Institute for Media Studies, 1994), 45.
31. Anne Hull, "Metal to Bone Day 3: Betrayals," *St. Petersburg Times*, 4 May 1993. Reprinted in *Best Newspaper Writing, 1994*, ed. Christopher Scanlan (St. Petersburg, Fla.: Poynter Institute for Media Studies, 1994), 67–68.

Acknowledgments

Andersen Windows. "Television? Are You Serious?" By permission of Andersen Windows and Campbell Mithun Esty Advertising.

Cadence. "Always Open" ad. © 1997 Cadence Design Systems, Inc. The Cadence logo is a trademark of Cadence Design Systems, Inc. Based on Edward Hopper's "Nighthawks" in the collection of The Art Institute of Chicago. Reprinted by permission of Cadence Design Systems, Inc.

Cadillac "Serve Hot" ad. "Catera™ The Caddy that Zigs." © 1997 GM Corp. All rights reserved. Reprinted by permission of General Motors Corporation.

CIBA Vision Corporation. "Foresight" Web site. © 1997 CIBA Vision. Reprinted by permission of CIBA Vision Corporation, a Novartis Company.

Columbia Sportswear Company. "One Possible Explanation" ad. Reprinted by permission of Columbia Sportswear Company and Borders Perrin and Norrander Inc.

Richard Ben Cramer. Excerpt taken from article on the Middle East from *The Philadelphia Inquirer*, March 19, 1978. Copyright © 1978 Richard Ben Cramer. Reprinted with permission of Sterling Lord Literistic, Inc. and The Philadelphia Inquirer.

Osborn Elliott. Excerpt taken from page 55 in *The World of Oz* by Osborn Elliott. Copyright © 1980 by Osborn Elliott. Used by permission of Viking Penguin, a division of Penguin Putnam, Inc.

Barton Gellman. Excerpt taken from an article on the assassination of Israeli Prime Minister Yitzhak Rabin, in *The Washington Post*, November 4, 1995. Copyright © 1995 The Washington Post. Reprinted with permission.

Billy Graham Evangelistic Association. The Billy Graham Twin Cities Crusade in 1996 ad from the St. Paul Pioneer Press, June 23, 1996. Reprinted with permission.

Bob Greene. Excerpt taken from "By Any Other Name." *Esquire*, September 31, 1981. Copyright © 1981 by Bob Greene. Reprinted by permission of Sterling Lord Literistic, Inc.

James Hardie Siding Products. "Frustrating Mother Nature" ad. © 1997 James Hardie Building Products Inc. All rights reserved. Reprinted by permission of James Hardie & Coy Pty Limited.

Harley-Davidson Motor Company. "Ch 16: The Hopelessly Addicted" ad. Courtesy of the Harley-Davidson Motor Company.

Anne Hull. Excerpt taken from "Metal to Bone Day 1: Click." *The St. Petersburg Times*, May 2, 1993. Reprinted by permission of the St. Petersburg Times.

Imation Corporation "feet first" ad. © 1996 Imation. All rights reserved. Reprinted by permission.

International Paper. "Cherries in Winter" ad. Reprinted by permission of International Paper.

Lauren Kessler and Duncan McDonald. Excerpt taken from *When Worlds Collide: A Media Writer's Guide to Grammar and Style*, Fourth Edition, pp. 137–148. Copyright © 1996 by Lauren Kessler and Duncan McDonald. Reprinted by permission of Wadsworth Publishing Company.

"Long Term Parking." Taken from *People*, May 25, 1998. Copyright © 1998 People, Inc. Reprinted by permission of People magazine. Photos: AP/Wide World Photos. All rights reserved. Reprinted by permission. Small insert photo of Rose Martin reprinted by permission of Mrs. Agatha St. Amour.

MilkPEP. "Where's *Your* Mustache?" ad. Reprinted by permission of the National Fluid Milk Processor Promotion Board.

The Missouri Group. Excerpt taken from pp. 14–15 in *News Reporting and Writing* by The Missouri Group. Copyright © 1996 by St. Martin's Press, Inc. Reprinted with the permission of St. Martin's Press, Inc.

Nike, Inc. "The Meek May Inherit the Earth" ad. Reprinted by permission of NIKE, INC.

Panasonic Consumer Electronics Company. "Take It to Extremes" Shockwaves Portable Audio ad. Reprinted by permission.

Partnership for a Drug-Free America. "Power of Grandma" ad. Reprinted by permission.

Polaroid Corporation. "Dog Pooh" ad. Chris Hooper, Art Director; Bob Kerstetter, Copywriter; Hunter Freeman, Photographer. Reprinted by permission.

Saturn Corporation. "Groceries" ad. © Saturn Corporation. Used with permission.

Simon MOA Management Company, Inc. "Mall of America" Parental Escort Policy statement, released Wednesday, September 4, 1996. Reprinted by permission.

Subaru Outback ad. Courtesy of Subaru of America.

Tom Wheeler. Excerpt from "When in Doubt, Ask!" © 1995 Fulcrum Publishing Inc., Golden, CO. All rights reserved. Reprinted from *Journalism Stories from the Real World*, Retta Blaney, Editor.

Toro Corporation. "Red White Blew" ad. Reprinted with permission.

Excerpt from *Wired* Magazine. Wednesday, June 25, 1997. Reprinted by permission.

Index